Environmental Policy and Management

SECOND EDITION

Environmental Policy and Management

EDITED BY
Younsung Kim
George Mason University

cognella®
SAN DIEGO

Bassim Hamadeh, CEO and Publisher
Peaches diPierro, Acquisitions Editor
Carrie Baarns, Manager, Revisions and Author Care
Kaela Martin, Senior Project Editor
Jeanine Rees, Production Editor
Jess Estrella, Senior Graphic Designer
Alexa Lucido, Licensing Manager
Natalie Piccotti, Director of Marketing
Kassie Graves, Senior Vice President, Editorial
Alia Bales, Director, Project Editorial and Production

Cover image copyright © 2023 iStockphoto LP/Sjo.

Printed in the United States of America.

cognella® | ACADEMIC PUBLISHING
3970 Sorrento Valley Blvd., Ste. 500, San Diego, CA 92121

CONTENTS

Preface   vii
Introduction   ix

UNIT I

Introduction to Environmental Policy, Climate Treaty, and Energy Transition   1

READING 1.1

Tragedy of the Commons and the U.S. Environmental Policy Approaches
Since 1970s   3
By Younsung Kim

READING 1.2

To and From the Paris Agreement: Review of Global Climate Governance
Toward Decarbonization   13
By Younsung Kim

READING 1.3

Energy Transition: From Fossil Fuels to Renewables for a Net-Zero,
Climate Resilient World   25
By Younsung Kim

UNIT II

Contemporary Issues and Toward a Sustainable World   37

READING 2.1

Public Participation and Environmental Justice: Access to Federal
Decision Making   41
By Dorothy M. Daley and Tony G. Reames

READING 2.2

Mushroom Packages: An Ecovative Approach in Packaging Industry   65
By Younsung Kim and Daniel Ruedy

READING 2.3

The Inclusive City: Urban Planning for Diversity and Social Cohesion   91
By Franziska Schreiber and Alexander Carius

READING 2.4

Renewable Energy Delivery and Expansion with Public and Private
Partnerships for the Global South   113
By Kyoo-Won Oh and Younsung Kim

READING 2.5

Carbon Markets: Principles and Current Practices   131
By Younsung Kim

PREFACE

Sustainability has emerged as a driving force for changing our modern society. It leads our society to shift to a different pathway that has never been before, pushing us to consider ecosystem value in our daily lives. Climate risk, food insecurity, and population growth, mostly in developing countries, document the inevitability of sustainable thinking and environmental policy considerations. It is not difficult to see the current system's perilous state and the urgent demands for a sustainable future.

We have a glimpse of hope in 2024, however. Some eventful changes include the rejoining of the Paris Agreement under the Biden administration in the United States and the adoptions of the 2021 bipartisan infrastructure bill and the 2022 Inflation Reduction Act in which electric vehicles and energy efficient products have been subsidized and energy transition towards renewables have been sped up. In parallel, other positive events were seen, such as Green New Deal proposals across the world, strong voices for correcting environmental injustice, climate capital and carbon pricing discussion in the European Union (EU), Environmental, Social, and Governance (ESG) Index-guided green investment opportunities, green funds created by the influential thinkers and leaders like Bill Gates and Jeff Bezos to fund clean breakthrough technologies, to name a few. Those trends and encouragement combined send us a strong signal to explore and educate ourselves concerning sustainability, via environmental policy.

As to environmental policy educational resources, we can easily find green information from more than 300,000 websites regarding climate change, blogs, and media. In terms of textbooks for students in higher education, we are fortunate to see many resourceful textbooks in the marketplace. Despite a wealth of information, I still found a gap among the many selections in that environmental policy textbooks are skewed to U.S. politics and environmental discussions that focused on problems. They elaborate on politics and conflicts, while solutions are less obviously discussed. Also, alternative policy tools such as private–public partnerships, clean innovation, carbon markets, and empowerment of citizen groups have not been adequately probed. Also, environmental justice has been embedded in the prior textbooks, rather than being discussed as a stand-alone theme.

As such, traditional environmental policy learning seems not fully aligned with the ways to cope with today's complicated, intertwined global environmental issues and create tomorrow's sustainable future.

This edition recognizes the chasm and carries on a more enriched understanding of currently available solutions to address and manage common problems in partnerships with diverse social actors. In doing so, this book may deepen and broaden the knowledge of the root causes of environmental policy and provide the direction to mitigate environmental impacts. It highlights three key points:

1. **Environmental issues are discussed with an interdisciplinary approach:** The articles in this anthology provide a multidisciplinary understanding of much complicated global environmental issues, introducing theoretical frameworks in environmental science, management science, and public policy as they relate to environmental policy issues.

2. **New policy instruments:** Articles in Unit II introduce the new trends in environmental policy and the emerging approaches to environmental policy (e.g., public-private partnerships, collaborative governance, etc.). A solution-based approach is also underscored. The emerging approaches will be useful in raising unanswered but important questions to be heard to advance environmental policy.

3. **Discussion questions that foster deep learning:** At the end of each chapter, you will find probing questions for discussion. They can be used to facilitate in-class discussions and promote self-learning.

In reading the articles, I wish you to find ingenuity, creativity, and hope for a sustainable future by better understanding the complicated nature of wicked environmental issues and learning about today's environmental treaties and tomorrow's technologies and approaches to our planet.

Lastly, I am indebted to my students in and outside of classrooms, from high school interns in my lab to undergraduates, graduate students, and adult learners across the world. They are all passionate about protecting the earth and their future from greedy capitalism and are prepared to undertake meaningful engagement in the policymaking process and implement substantive actions. They helped me become a better educator.

This book is dedicated to innovative thinkers and inspiring leaders for our own planet.

Younsung Kim
McLean, Virginia

INTRODUCTION

This edited book is composed of two units. The main themes of the two units are as follows.

Unit I covers the fundamental principles of environmental policy and politics. It focuses on these guiding questions:

- What are the main features of environmental policy?
- How is environmental policy different from public policy?
- How is policy formed?
- Who are the key actors in making and implementing (environmental) policies?
- What was the distinct approach taken at the early stage of the U.S. modern environmental policy?
- How have we (U.S. Americans) arrived at adopting environmental laws to address industrial pollution and manage natural resources more wisely?

Unit I also outlines environmental issues beyond geographic boundaries and not limited to the natural/ecological realm. Two issues are highlighted: climate change and energy. They are considered as not only environmental but political, economic, and technical challenges. The solutions for each transboundary challenge are investigated. The guiding questions are:

- What are the climate change policy approaches that have been considered?
- Which carbon control policies are better?
- Why is the 2015 Paris Agreement considered to lead the monumental changes in international climate politics?
- What are the major challenges for renewable-based electricity?
- What are the governmental actions to promote renewable-based electricity?

Unit II discusses the emerging environmental issues and policy approaches. Indeed, the issue of environmental justice (EJ) has long been reported, but systematic inequality in environmental decision making has not yet been adequately addressed. Other chapters investigate the key practices that are assumed to promote sustainability in the context of waste management, sustainable business, and urban resilience. The unit focuses on these guiding questions:

- How has environmental inequality been laid out in the United States based on people's socio-economic status and races?
- What are the major challenges for managing waste these days? How could we promote waste management more radically?
- How has the U.S. federal government attempted to address the issue? Did it work?
- How can we make a business sustainable? Are the firms claiming that they are sustainable trustworthy?
- What is the concept of a sustainable city? What would be policy approaches to make cities sustainable?
- How can we increase energy access in developing countries, particularly in rural areas of countries?
- What is a successful example of improving solar-based energy access that does not harm the environment?
- What is the concept of sustainability? What are the key challenges of sustainability down the road at the end of the 21st century? Which policy approaches should be more emphasized? Which actors need to be (more) empowered?
- What are the principles of carbon markets, and how many carbon pricing schemes have been taken in place to remove carbon emissions?
- How have carbon markets shown variances in unit prices and approaches to include greenhouse gases?

Introduction to Environmental Policy, Climate Treaty, and Energy Transition

U nit 1 contains three chapters. The first chapter introduces the Tragedy of the Commons terminology and discusses how natural resource mismanagement and industrial pollution are tied to the tragic state of our planet. The unit also outlines the policymaking cycle, which includes agenda setting, option formulation, decision making, policy implementation, and evaluation. The cyclic approach offers a needed insight into actors that could be influential in each stage, as well as the political implications related to policy changes and policies that are favored in a status quo.

In the second chapter, climate change is referred to as the biggest market failure. As a global negative externality case, climate change needs to be addressed urgently, and its physical impacts and risks are evident in ecosystems and human society. International policymakers have responded to climate change by establishing institutional frameworks and modifying them through multilateral meetings. The 1992 UNFCCC, 1997 Kyoto Protocol, and 2015 Paris Climate Deal are zoomed in to characterize the tectonic shifts in rules and discourses in response to climate change. The 1992 UNFCCC aimed to stabilize global greenhouse gas emissions and suggested Conference of the Parties meetings to show progress toward the carbon reduction commitment. The Kyoto Protocol created market-based mechanisms for climate mitigation and required 38 developed countries to reduce greenhouse gas emissions. The 2015 Paris Climate Agreement scaled up climate mitigation and adaptation measures by emphasizing carbon financing and collaborative partnerships with multi-sectors, asking each country to submit Nationally Determined Contributions (NDCs) to confirm their climate action plans and implementations. The chapter presents a critical outlook on why we should maintain the agreement and create more diverse approaches to push all institutional actors and individuals to adopt carbon-reducing measures so that they can align with the Paris goal of limiting the global temperature increase to well below 2 degrees Celsius above pre-industrial levels.

The final chapter outlines the shift in energy policy from reliance on fossil fuels to a preference for renewable sources. It introduces Renewable Portfolio Standards (RPS) and Feed-in-Tariffs (FITs) as the main policy instruments utilized in the United States and European Union. RPS has been instrumental in encouraging the uptake of renewable electricity in some U.S. states, in the absence of a robust national renewable energy strategy. Meanwhile, FITs have provided incentives for renewable energy project developers in leading renewable nations such as Germany and Spain, encouraging the supply of electricity from renewable sources. Additionally, the chapter covers the most recent policy developments for energy transition in the United States, European Union, and China, and introduces the emerging policy concept of a just transition. The U.S. Department of Energy's Office of State and Community Energy Programs (SCEP), under the current Biden administration, is highlighted as an example of how to implement a just transition in practice and accelerate the energy transition.

Tragedy of the Commons and the U.S. Environmental Policy Approaches Since 1970s

By Younsung Kim

S ince the 1970s, environmental policy in the United States has noticeably evolved to address deteriorating environmental quality and public health concerns. Major laws to regulate air pollution, water pollution, and waste issues were passed mostly in the 1970s, and the Environmental Protection Agency (EPA) was created to implement environmental laws in partnership with state and local governments. Policies for the conservation of nature also changed profoundly during the modern environmental policy era. As a flagship legislative change, the Endangered Species Act was greatly strengthened in 1973 to identify and list endangered or threatened species of plants and animals. The U.S. environmental quality has shown substantial progress owing to environmental policy drives and public support (Kraft and Vig, 2022).

Despite its overall success, we still face a wide array of challenges surrounding our ecosystems, existing and new. For instance, the U.S. water quality shows relatively modest improvement compared to the air quality (US EPA, 2017). Also, continuously strengthening evidence on climate change and biodiversity crisis indicates the more accelerated speed and scale of governmental intervention, but passing environmental laws is getting more difficult due to a partisan divide and issue polarization (Kraft and Vig, 2022). Understanding the nature of environmental problems and the policymaking process can present an insight into why policies have not been passed successfully. It will also hint at better policy approaches to complicated, transboundary environmental issues, including climate change. This chapter aims to outline the unique characteristics of environmental problems, the policymaking process, and major policy approaches taken over the past five decades. It also presents the merge of environmental and climate issues nowadays, chronicling the pushes and pulls of climate policies since the Obama administration.

Tragedy of the Commons: Two Types

The Tragedy of the Commons is a foundational concept to make it easier to understand a class of environmental problems. The concept originated from an article by ecologist and biologist Garret

Hardin published in *Science* (Hardin, 1968). In his article, Hardin indicated that environmental problems belong to a class of no technical solutions, underscoring the insolubility without reducing inputs of pollution to the environment as well as adopting adequate governmental interventions relying on coercive policy mechanisms. The premise of the concept is that humans are rational and are prescient of how to maximize their welfare. Many environmental problems arise due to humans' self-interest maximizing motives that result in excessive use of natural resources and environmental goods when there is a lack of an institutional system to curb human desires. Natural resources have been overexploited, and industrial pollution has been prevalent accordingly.

Under the Tragedy of the Commons analogy, the environmental problem is largely classified into two types. The first type of tragic status in commons is seen when humans have taken something good out of commons. Suppose a herdsman raises cattle in an open pasture. Each herdsman is incentivized to put as many cattle as possible in the commons if there are no limits to adding cattle and the benefits of more cattle are greater than the costs. Putting more cattle is then a rational choice since everyone in the commons would share the costs of additional cows raised. As such, nearby streams of cattle-ranching areas would be polluted, and the soil would get infertile. Deforestation, desertification, and overfishing exemplify this kind of the Tragedy of the Commons, which is the overuse and mismanagement of natural resources.

In another case, the Tragedy of the Commons occurs when something bad is put into the commons. On this occasion, people are driven to emit toxic and noxious gases into the air, discharge harmful water pollutants into freshwater, and dump hazardous chemicals and wastes on land where no regulatory guidelines have been adopted. As such, the benefits of pollution surpass the costs of treating pollutants. Industrial pollution without regulations brings ruin to all, creating a tragic status in the commons. It incentivizes people not to pay for their pollution and induces costs borne by society. As the second type of Tragedy of the Commons, industrial pollution illustrates the high costs of negative externality, which is the spillover effect of transactions when someone's action negatively affects others' welfare.

Those two types of the Tragedy of the Commons cannot be addressed without proper governmental intervention and justify the need for establishing a set of environmental policies with the polluters-pay-principle (PPP). The is the commonly accepted practice that those who produce pollution should bear the costs of managing it to prevent damage to human health or the environment, and it is guided by the 1992 Rio Declaration as part of a set of broader principles to promote sustainable development worldwide. Well-designed environmental policies would be capable of correcting the pricing systems of undervalued natural resources and environmental goods and services (Olmstead, 2022).

Policymaking Process

Making environmental policy to establish coercive regulatory mechanisms in which people find it cheaper to control the overuse of natural resources and pollution is not straightforward. This is in part due to environmental policymaking requiring the coordination of multiple actors whose stakes vary across the policymaking process under high-level scientific and political uncertainties (Young and Stokke, 2020).

Policy formulation requires an understanding of the policy process. Rosenbaum (2020) defines policymaking as a process that involves a number of related decisions originating from different institutions and actors ranging across the whole domain of the federal government and private institutions. This process involves the five different steps from agenda-setting, option formulation, decision-making, and policy implementation to evaluation. At the stage of policy formulation, a policy agenda is set, and alternative options to solve the policy issues are actively discussed. This is the stage where environmental science, economics, and policy analysis all contribute to the development of a policy agenda. U.S. presidents and the executive branches of the government have played a critical role in agenda-setting, as seen in the latest case of President Obama with his climate-related achievements, including the Clean Power Plan (Vig, 2022).

Policy legitimation is a stage of selecting and endorsing policies through political actions by Congress, the president, and courts. Political support for formal enactment needs to be mobilized during the decision-making stage, and many policy proposals tend to fail due to polarized public and partisan divides. Indeed, President Biden's Build Back Better legislative agenda, which includes ambitious climate actions costing about $555 billion over 10 years, couldn't successfully pass the Congress for such reasons (Leonhardt, 2022).

Policy evaluation can create an agenda, reflecting on the performance of certain policies and programs adopted. At the stage of policy evaluation, government agencies, citizen groups, environmental think thank groups having research arms, or Congressional Research Services (CRS) play a vital role in analyzing governmental programs with one or multiple criteria, such as health and ecological considerations, cost efficiency, and moral imperatives (Greenberg, 2007).

Figure 1.1.1 presents the five steps of policymaking from agenda-setting to policy evaluation. Once a policy is evaluated, the analysis outcomes can be fed into the agenda-setting stage, creating a cyclic model of policymaking.

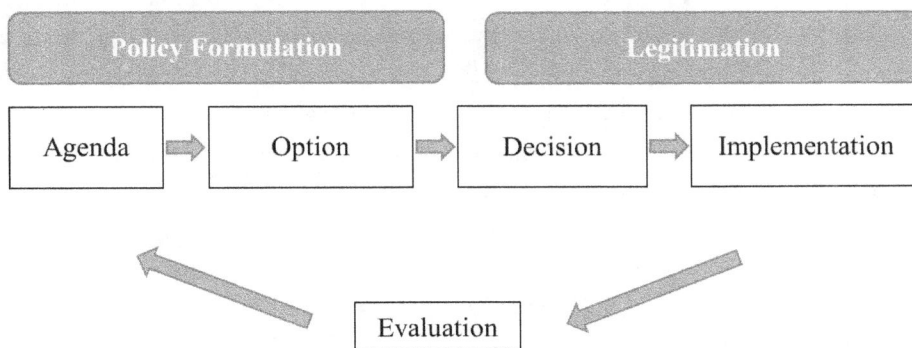

Figure 1.1.1 The Five Stages of Policymaking Process, from Agenda-Setting, Option Formulation, Decision, Policy Implementation, and Policy Evaluation.

Uniqueness of Environmental Issues Creating Conflicts

Environmental politics occur when stakeholders' interests and values for the environment are varied and clash during decision-making. There are two critical features of U.S. environmental policymaking. First, environmental policy conflicts almost concern fundamental differences in values. In most cases, people's values are divided into two groups, environmentalists and cornucopians. Environmentalists acknowledge the pristine value of nature and promote preservation in one spectrum. They also value conservation for recreation and advocate the prudent use of natural resources (Layzer, 2012). Cornucopians believe in humans' ingenuity to invent and engineer solutions to solve natural resource scarcity and pollution as technologies advance. As technology optimists, they tend to place immense value on individual liberty and criticize environmental regulatory policies that would cause unnecessary burdens to society and industry alike. Those two groups converse and are in conflict in every decision regarding public land management, adoption of new environmental policies, and strengthening of environmental regulations.

Also, environmental conflict occurs due to the way problems are defined and the solutions depicted can shape how those values get translated into policies. Environmental issue framing is reliant on the uses of environmental science and environmental economics concepts. Many government agencies, including the U.S. EPA, rely on sound science in the belief that the use of scientific information in policymaking can lead to *better* natural resource management decisions and more *effective* environmental policy. However, scientific uncertainty impedes the use of scientific evidence for environmental decision-making. Driven by restricted data, inconclusive outcomes, and measurement error, scientific uncertainty can be reduced but cannot be eliminated. Climate science illustrates how climate deniers use scientific uncertainty to advocate their argument against climate policies. The Intergovernmental Panel on Climate Change (IPCC) has thus updated its assessment reports and sought to reduce scientific uncertainty despite its doubt over what difference more reports will make.

Environmental economics is another competing tool to frame environmental issues and policies in the context of benefits and costs. As environmental policies try to address the issues of mispriced environmental goods and services, environmental protection always comes at a cost. The framing based on projected benefits and costs raises several questions that are hard to be answered. Some questions would be the following: who will bear the costs, who will get benefited from environmental protection, whether the benefits of environmental protection are greater than its costs, how to minimize regulatory costs, how to measure the benefits of environmental regulations, and whether or not to adopt regulations with negative social costs is morally acceptable.

Indeed, climate-related policy conflicts demonstrate a prime example of why federal policies with broader impacts often fail to be adopted and implemented. Mounting evidence of climate science (IPCC, 2022; Plumer and Zhong, 2022) and climate risks to humans and ecosystems, particularly coastal communities (Crimmins et al., 2016) only reassure urgent, immediate, and impactful actions to reduce greenhouse gases (GHGs).

U.S. Environmental Policy Approaches

The early stage of the U.S. environmental policy was deterrence-based, as policy responses were led by increasing environmental crises and public outcry over the right to live in an amenable environment. The nation thus relied on business and industry-prescribed environmental protection methods of emissions reduction and technologies like catalytic converters or scrubbers to filter sulfur dioxides. In this approach, the government relied on strict timelines for compliance, and expensive enforcement mechanisms, including audits and penalties, were accompanied. This traditional approach is called command-and-control, top-down, or direct regulation, and the Clean Air Act of 1970 and the Clean Water Act of 1972 set examples. This approach, in part, worked well as most environmental problems back then were local, and not much information was known about pollution abatement costs (Ellerman, 2007; Stavins, 2007).

However, the industry under the command and control approach was less cooperative and didn't comply with the regulations, and legal suits were used as a tactic to avoid or reduce penalties or delay payment. Pollution abatement costs were immense in both industry and society, creating a strong demand for smart regulations in the 1980s and 1990s. With the pre-industrial mantra, the Reagan administration pursued the balancing act of costs and benefits associated with environmental protection. The second generation of environmental policy approach underscoring market-based economic incentives has emerged. Under this approach, the flexibility of how to meet the regulatory standards has become the greatest concern, and policies adopted need to be cost-effective to ensure pollution control with the cheapest options. The approach considers pollution-reducing entities' different capacities for pollution control, and allowance-trading programs, pollution taxes, and subsidies for environmentally friendly products or behaviors were highly encouraged. As such, this flexible approach allows businesses to select their own cost-effective strategies for reducing emissions (Press and Mazmanian, 2018).

On top of this approach, the United States has relied on business volunteerism, wherein businesses commit to environmental and sustainability goals that exceed those required. They oftentimes get exemptions from governments' enforcement actions or any mandatory regulatory approaches. Under the Bush administration, tackling climate change has been hinged upon firms' voluntary commitments (Kim and Darnall, 2016) and actions for carbon reducing actions, and some states' progressive environmental activism (Rabe, 2022). The environmental policy shift from a command-to-control approach to volunteerism illustrates the nature of environmental issues not limited to an environmental medium or species but tracing across ecosystems and the planet.

Table 1.1.1 presents the changes in regulatory approaches from a traditional, command-and-control regulatory approach to volunteerism and policy examples under each approach.

Table 1.1.1 The emergence of the U.S. environmental regulatory approaches since 1970s

Time Period	Environmental Policy Approach	Examples	Features
1970s~1980s	**Command-and-control**	• Clean Air Act of 1970 • Clean Water Act of 1972 • Resource Conservation and Recovery Act of 1976	• Strict timeline • Permit-based approach relying on the polluter-pays principle • Heavy penalties • Enforcement mechanisms
1980s~2000s	**Market-based economic incentives**	• Sulfur dioxide (SO_2) allowance-trading program under the Clean Air Act Amendments of 1990	• Flexibility of how to meet policy goals • Priced-based or quantity-based policy goal • Uncertain of prices or quantities of pollution reduced
2000s~	**Volunteerism**	• US EPA Performance Track Program • US DOE and EPA's Energy Star	• Promoting eco-labels • Relying on firms' self-regulatory approaches • Process or product-based approaches to distinguishing environmental quality from conventional products or production practices

Embracing Carbon Neutrality and Next Steps

The latest IPCC six assessment report witnesses the climate impacts are already more widespread and severe than expected. Droughts, extreme heat, and record floods already threaten food security and livelihoods for millions of people. Since 2008, floods and storms have forced more than 20 million people from their homes each year. The report underscores the irreversible impacts of climate change. (IPCC, 2022; Plumer and Zhong, 2022). The United States has also recognized climate change as a serious threat to human health and well-being. For instance, coastal communities face greater vulnerability to health impacts from flooding (Crimmins et al., 2016).

Nowadays, environmental policy has been integrated into climate policies, and under the Obama administration, the United States has resumed its commitment to the international climate treaty and seemed to regain its leadership for the 2015 Paris Climate Deal. As a national policy goal, the United States during the Obama administration pledged to curb emissions between 26% and 28% below 2005 levels by 2025, with a longer-term goal of an 80% reduction by 2050. Under this goal, the Clean Power Plan was adopted to reduce GHG emissions from power utilities, and its goal was to reduce carbon pollution from the power sector by 32% of emissions in 2005. Fossil fuel–fired power plants are by far the largest source of U.S. carbon emissions, making up 31% of total GHG emissions (U.S. EPA, 2015). Also, Corporate Average Fuel Economy (CAFÉ) standards were strengthened to reduce GHG emissions from the transportation sector, and the most ambitious CAFÉ standard was to increase fuel economy to the equivalent of 54.5 mpg for cars and light-duty trucks by model year 2025 (U.S. EPA, 2012). However, these rules were overturned by the Trump administration, demonstrating the instability and reversibility of federal climate policies if laws were not enacted.

Due to the Biden administration's resuscitating push for climate agendas, the United States has rejoined the Paris climate treaty and attempted to restore credibility. President Biden has mapped out a $2.2 trillion clean energy and green jobs plan to cut emissions from electricity to zero by 2035. The plan is assumed to help the United States achieve net-zero emissions by 2050. However, the inclusion of social policies with climate change programs made the Build Better Back Act fail to pass the Senate. Conceived is a critical insight into a political strategy that would focus on key programs once, rather than clumping all issues together in a proposed legislative bill (Dayen, 2021).

However, the EU has successfully passed the Green New Deal in 2019 and placed carbon neutrality as the center of green economy policies. Its momentum has not dwindled even under the global pandemic crisis. The corporate sector, multilateral banking agencies, central banks, and local governments worldwide and in the United States have aligned with such 2050 carbon-neutral initiatives. The worldwide trend and agenda for zero-carbon society questions how the U.S. federal policies could sustainably push forward net-zero carbon emissions, which policy directives need to be pursued, and how the federal government could solidify its collaboration with the private sector, civil society, states, and local governments. And finally, the country's leadership in the international climate regimes needs to be continually conceived in alliance with the national security concern and future economic opportunities.

References

Crimmins, A., Balbus, J., Gamble, L., Beard, C. B., Bell, J. E., Dodgen, D., Eisen, R. J., Fann, N., Hawkins, M., Herring, S. C., Jantarasami, L., Mills, D. M. Saha, S., Sarofim, M. C., Trtanj, J., & Ziska, L. (2016). Executive summary. The impacts of climate change on human health in the United States: A scientific assessment. U.S. Global Change Research Program, Washington, DC, 24 pp. http://dx.doi.org/10.7930/J00P0WXS

Dayen, D. (2021, October 26). How to fix the democrats' build back better plan. *The New York Times.* https://www.nytimes.com/article/build-back-better-explained.html

Ellerman, A. D. (2007). Are cap-and-trade programs more environmentally effective than conventional regulation? In J. Freeman & C. D. Kolstad (Eds.), *Moving to markets in environmental regulation* (pp. 48–59). Oxford University Press.

Greenberg, M. R. (2007). Introduction: A quick walk through a framework of six environmental policy criteria. In *Environmental Policy Analysis and Practice* (pp. 1–13). Rutgers University Press.

Hardin, G. (1968). The tragedy of the commons. *Science, 162,* 1243–1248.

IPCC (2022). *Climate change 2022: Mitigation of climate change. Contribution of Working Group III to the Sixth Assessment Report of the Intergovernmental Panel on Climate Change.* In P. R. Shukla, J. Skea, R. Slade, A. Al Khourdajie, R. van Diemen, D. McCollum, M. Pathak, S. Some, P. Vyas, R. Fradera, M. Belkacemi, A. Hasija, G. Lisboa, S. Luz, & J. Malley (Eds.). Cambridge University Press. doi 10.1017/9781009157926.001

Kim, Y., & Darnall, N. (2016). Business as collaborative partner: Understanding firms' sociopolitical support for policy formation. *Public Administration Review, 76*(2), 326–337.

Kraft, M. E., & Vig, N. J. (2022). US environmental policy: A half-century assessment. In N. J. Vig, M. E. Kraft, & B. G. Rabe (Eds.), *Environmental policy: New directions for the twenty-first century* (pp. 3–33). CQ Press.

Layzer, J. (2012). A policymaking framework: Defining problems and portraying solutions in U.S. environmental politics. In *The environmental case: Translating values into policy* (pp. 1–21) (3rd ed.). CQ Press.

Leonhardt, D. (2022). Will climate action happen now: More democrats are focusing on it? https://www.nytimes.com/2022/01/21/briefing/climate-change-bill-democrats.html

Olmstead, S. (2022). Applying market principles to environmental policy. In N. J. Vig, M. E. Kraft, & B. G. Rabe (Eds.), *Environmental policy: New directions for the twenty-first century* (pp. 227–248). CQ Press.

Plumer, B., & Zhong, R. (2022). Stopping climate change is doable, but time is short, U. N. panel warns. *The New York Times.* https://www.nytimes.com/2022/04/04/climate/climate-change-ipcc-un.html

Press, D., & Mazmanian, D. A. (2018). Toward sustainable production: Finding workable strategies for government and industry. In N. J. Vig, M. E. Kraft, B. G. Rabe (Eds.), *Environmental policy: New directions for the twenty-first century* (pp. 269–274). CQ Press.

Rabe, B. (2022). Racing to the top, the bottom, or the middle of the pack? The evolving state government role in environmental protection. In N. J. Vig, M. E. Kraft, B. G. Rabe (Eds.), *Environmental policy: New directions for the twenty-first century* (pp. 35–62). CQ Press.

Rosenbaum, W. A. (2020). Making policy: The process. In W. A. Rosenbaum, *Environmental politics and policy* (pp 30–61). CQ Press.

Stavins, R. (2007). Market-based environmental policies: What can we learn from U.S. experience (and related research)? In J. Freeman & C. D. Kolstad (Eds.), *Moving to markets in environmental regulation* (pp. 19–47). Oxford University Press.

U.S. EPA (2012, August). EPA and NHTSA set standards to reduce greenhouse gases and improve fuel economy for model years 2017–2025 cars and light trucks. EPA-420-F-12-051.

U.S. EPA. (2015). Overview of the clean power plan: Cutting carbon pollution from power plants. https://archive.epa.gov/epa/sites/production/files/2015-08/documents/fs-cpp-overview.pdf

U.S. EPA. (2017). National water quality inventory: Report to Congress. EPA 841-R-16-011. https://www.epa.gov/sites/default/files/2017-12/documents/305brtc_finalowow_08302017.pdf

Vig, N. J. (2022). Presidential powers and environmental policy, In N. J. Vig, M. E. Kraft, & B. G. Rabe (Eds.), *Environmental policy: New directions for the twenty-first century* (pp. 87–110). CQ Press.

Young, O. R., & Stokke, O. S. (2020). Why is it hard to solve environmental problems? The perils of institutional reductionism and institutional overload. *International Environmental Agreements, 20*, 5–19. https://doi.org/10.1007/s10784-020-09468-6

Zhong, R. (2022). These climate scientists are fed up and ready to go on strike. *The New York Times.* https://www.nytimes.com/2022/03/01/climate/ipcc-climate-scientists-strike.html

DISCUSSION QUESTIONS

1. How does the Tragedy of the Commons analogy relate to climate change?

2. There are policy actors involved in the policymaking process inside and outside the government. In which stage of policymaking would courts be actively engaging in and shaping environmental policy?

3. Policymaking is viewed as a cyclic model, and what are the implications of the cyclic nature of policymaking in the U.S. political system?

4. What are the strengths and limitations of volunteerism-based environmental policy approaches? Would the approach be recommendable for environmental issues with higher health risks?

5. What are the greatest hurdles to achieving the U.S.'s carbon-neutral goals by 2050? Which practical and concrete programs have been adopted and proven successful, if any?

To and From the Paris Agreement

Review of Global Climate Governance Toward Decarbonization

By Younsung Kim

Introduction

The ever-increasing climate scientific evidence has presented an urgent and strong need for global climate policies. In 2015, 195 countries and the European Union adopted the Paris Agreement (PA) after a decade of negotiations under the United Nations Framework Convention on Climate Change (UNFCCC). Many policymakers and environmental advocates consider the agreement a historic milestone for climate policy, crediting its success to extensive bilateral and multilateral channels to dialogue and agree on climate actions. However, most governments have shown slow progress, and the possibility of limiting global temperature increases to 1.5 degrees Celsius seems unrealistic (Mooney, 2023). Still, it is evident that the PA has played a critical role in encouraging nations worldwide to work towards addressing climate change.

This chapter intends to sketch out the landscape of global climate treaties, focusing on regulatory systems. It focuses on three major multilateral agreements, the 1992 UNFCCC, the 1997 Kyoto Protocol, and the 2015 Paris Agreement. The PA is particularly underscored, as it currently guides countries in developing carbon-reduction targets and initiatives while encouraging all social and economic actors to work towards decarbonization. First, the PA has several key features that attempt to overcome the limitations of the former climate treaty. Reflective of the increasing greenhouse gas (GHG) emissions concentrations trends in developing countries such as China, the PA contains collective commitments to carbon reductions by both advanced economies and developing nations. Second, agreements regarding climate adaptation and finance have been added to the PA (Bodle et al., 2016). Those unique aspects of the PA will be described. Last, reducing emissions from deforestation and forest degradation (REDD+) enshrined in Article 5 of the PA has gained attraction from international policymakers, project developers, and environmental advocates. REDD+ has the potential to reduce GHG emissions while minimizing deforestation in biodiversity hotspots in developing countries. The mechanisms, benefits, and limitations of REDD+ are presented in this chapter.

Global Climate Agreements: UNFCCC and Kyoto Protocol

In 1990, the Intergovernmental Panel on Climate Change (IPCC) published the first assessment report, acknowledging that the earth has been warmed due to increasing anthropogenic GHG emissions. In response, the United Nations Framework Convention on Climate Change (UNFCCC) was adopted in 1992, setting a broad strategy for countries to work jointly to address climate change. The long-term objective was to *stabilize* global GHG concentrations in the atmosphere at a level that would prevent dangerous anthropogenic interference with the climate system (Article 2). It was assumed that global anthropogenic carbon dioxide (CO_2) emissions could be stabilized at levels ranging between 350 and 750 parts per million (ppm) by around the year 2100 (IPCC, 1996). However, in 2022, worldwide carbon emissions data present 420 ppm[1] with an increasing trend, which is highly above from 280 ppm, the approximate historical CO_2 concentrations for up to 800,000 years prior to the industrial revolution (Jordan et al., 2015).

UNFCCC relies on the "common but differentiated responsibilities" principle, which shaped the evolution of the climate regime and has played an important role in promoting compromise and agreement (Brunnée & Streck, 2013). Under the principle, all countries share an obligation to act. Still, industrialized countries are responsible for leading climate mitigation due to their relative wealth and disproportionate contribution to the problem through historic emissions. Forty countries and the European Union are listed in Annex I of the UNFCCC, and they agreed to reduce their emissions to the 1990 level, but the timeline for carbon reductions has not yet been fixed. The UNFCCC agreed that member states would undertake climate research. Also, member states decided to meet at Conference of the Parties (COPs) to discuss climate mitigation and adaptation issues and countries' detailed regulatory commitments (Selin & VanDeveer, 2022).

The second multilateral climate agreement is the Kyoto Protocol. In 1997, the Kyoto Protocol was adopted amidst the growing scientific evidence presented in the second IPCC report and its concerns over increasing social and environmental impacts. In recognition of climate risks, Annex I countries committed to reducing their GHG emissions collectively by 5.2% below 1990 levels by 2008–2012 (Selin & VanDeveer, 2022). The targets for the first commitment period of the Kyoto Protocol cover emissions of the six major GHGs, which are carbon dioxide (CO_2), methane (CH_4), nitrous oxide (N_2O), hydrofluorocarbons (HFCs), perfluorocarbons (PFCs), and sulfur hexafluoride (SF_6). Following the common but differentiated responsibilities, individual countries' emission reduction targets were set differently. For instance, the United States and Canada committed to reducing GHGs by 7% of their 1990 baseline GHGs, while the EU-15 decided to lower GHGs by 8%. Interestingly, Norway could increase its GHGs by 1%, while Russia did not need to reduce its GHGs.

The Kyoto Protocol also laid out mechanisms to meet GHG emissions reduction targets. First, countries could reduce their domestic GHG emissions by relying on national policies or initiatives to lower carbon emissions. For example, countries could preserve forests or adopt sustainable practices

---

1 Atmospheric CO_2 concentration data is measured at the Mauna Loa Observatory, Hawaii (NOAA). CO_2.earth releases daily CO_2 and monthly averaged CO_2 concentration data (CO_2.earth, 2023)

such as energy efficiency improvement or reforestation. Second, emissions could be reduced more flexibly by Kyoto mechanisms. The Kyoto mechanisms allow GHGs to be reduced more cost-effectively by three market-based mechanisms—Clean Development Mechanism (CDM), Joint Implementation (JI), and International Emissions Trading (IET). CDM is a project-based carbon reduction, involving investment in emissions reduction or removal enhancement projects in developing counties and contributing to sustainable development. JI takes the same approach as CDM but enables developed countries to carry out emission reduction or removal enhancement projects in other developed countries or economies in transition, such as Russia or Ukraine. Modeling after the U.S. sulfur dioxides allowance trading mechanism, IET enables country-to-country trades of reduced carbon emissions (Green, 2021). While IET had not been taking place, the European Union launched the EU Emissions Trading System (ETS) in an attempt to meet the EU's emission target mandated by the Kyoto Protocol (Kim, 2023).

Paris Agreement (PA): A Landmark Achievement

Although the Kyoto Protocol entered into force in 2005, GHG emissions, including naturally existing greenhouses such as carbon dioxide, methane, and nitrous oxide, continued to rise. The 2009 UNFCCC conference, known as the Copenhagen Summit, aimed to establish a more effective successor to the Kyoto Protocol, but it was unsuccessful, leading many people to doubt the sustenance of multilateral climate diplomacy (Victor, 2011). Six years later, the 2015 PA turned over such doubt, finally ending long-standing UN-sponsored negotiations and marking it as a remarkable achievement (Falkner, 2016).

The new climate agreement aims to limit global temperature rise to below 2 degrees Celsius, and even 1.5 degrees Celsius, to improve our ability for climate adaptation. That involves making financial flows consistent with low-emission and climate-resilient development. Some key issues, such as technology, capacity building, reporting, and accounting, are detailed in PA articles. As an assessment tool, the global stocktake (Article 14) was also embraced to evaluate parties' collective efforts toward the treaty's goals (Bodle et al., 2016). Like taking inventory, global stocktake examines every aspect of global climate action and support, identifying the gaps, and working together to agree on solutions pathways to 2030 and beyond. In light of the stocktake's outcomes, parties are expected to strive for "progression" beyond previous efforts (Bodle et al., 2016).

Under the PA, each country has the autonomy to establish its targets and commitments. These targets and commitments are known as Nationally Determined Contributions (NDCs) and must be updated every five years. This forms a "pledge and review" system in which each party submits updated NDCs to the UNFCCC. However, the COVID-19 pandemic caused a delay in the NDC update process. In 2020, all parties needed to submit their updated NDCs, but the process was delayed as the 26th COP of the UNFCCC was postponed.

The PA is considered a significant milestone towards carbon neutrality as it tends to harness nonstate actors' willingness and enthusiasm about climate actions. In this light, global climate governance is becoming polycentric, and carbon reductions or climate adaptation occur at multiple levels of authority and in various forums beyond the UNFCCC (Falkner, 2016). Despite the slow progress by states, regional, or local governments, private sectors, local communities, and nonprofit organizations strongly support a paradigm shift toward economic growth with GHG reductions and step in for progressive climate actions.

The success of the PA also depends on the public. It is important to emphasize individual actors' ability to adapt and align with the PA (Kim & Mackey, 2014), and environmental treaties no longer represent static contractual agreements among states (Gehring, 2007). It is also rather dynamic institutional arrangements that establish ongoing regulatory or legislative processes (Gehring, 2007) and act as lawmakers (Brunnée, 2012). As a result, in most international environmental regimes, the treaty text represents only the surface level of the norms (Bodansky et al., 2007). Most of the norms are adopted through relatively flexible and dynamic processes, which provide the system with adaptive capacity. The success of the PA will thus depend on implementation by all polycentric actors.

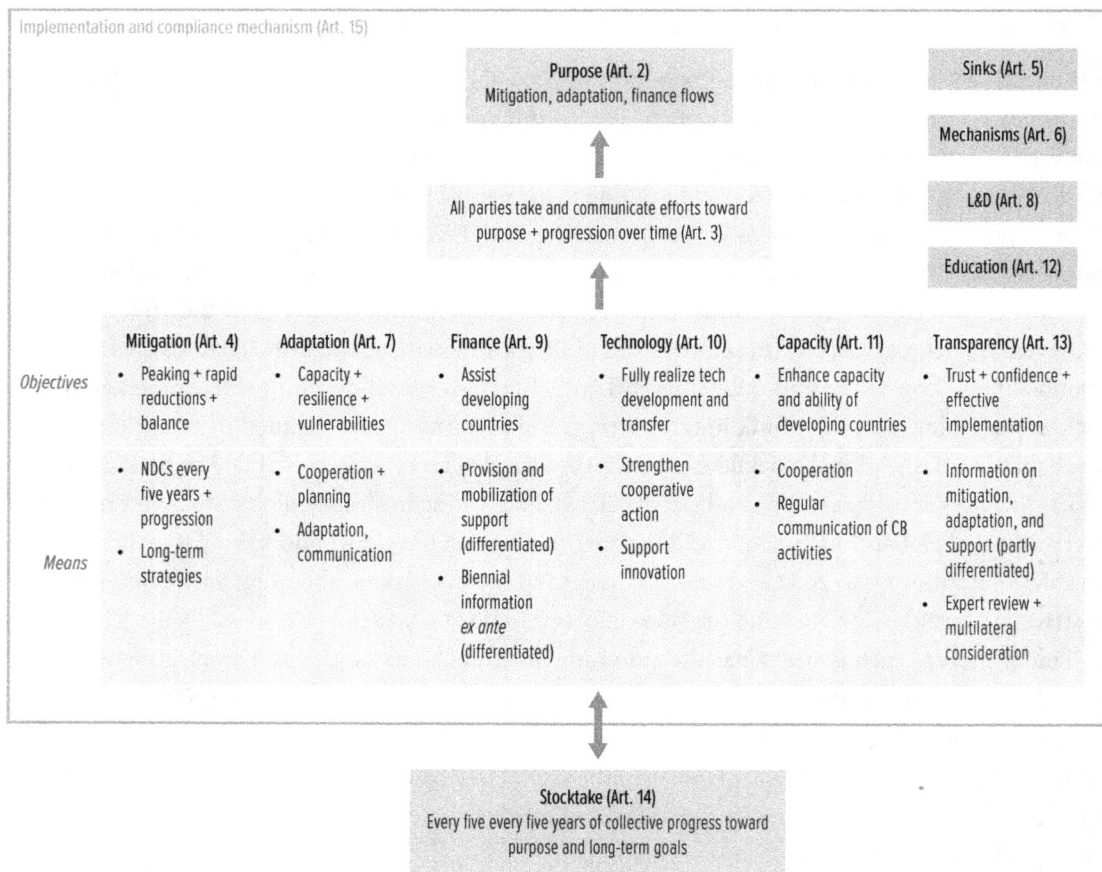

Figure 1.2.1 Structure of key issues in the Paris Agreement.
Source: Ralph Bodle, Lena Donat and Matthias Duwe, "Structure of Key Issues in the Paris Agreement from 'The Paris Agreement Analysis, Assessment and Outlook,'" *Carbon Climate Law Review*, vol. 10, no. 1, pp. 7. Copyright © 2016 by Lexxion Verlagsgesellschaft mbH.

	2015	2016	2017	2018	2019	2020	2021	2022	2023	2024	2025
NDC submission	INDC (2020–2025/30) submission				NDC preparation	NDC (2025–2030/35) submission				NDC preparation	NDC (2030–2035/40) submission
Secretariat						Synthesis report of NDCs					Synthesis report of NDCs
Stocktakes				Facilitative dialogue					First global stocktake		

Figure 1.2.2 Timeline for the NDC cycle.
Source: Ralph Bodle, Lena Donat and Matthias Duwe, "Timeline for the NDC Cycle from 'The Paris Agreement Analysis, Assessment and Outlook,'" *Carbon Climate Law Review*, vol. 10, no. 1, pp. 10. Copyright © 2016 by Lexxion Verlagsgesellschaft mbH.

Climate Finance: Key to Meeting the PA Goal

Climate finance is considered a crucial element in addressing climate change. It enables the funding of mitigation efforts, such as the transition from fossil fuels to renewable energy sources, and adaptation projects, like building climate-resilient infrastructure. These efforts require significant financial resources, and climate finance facilitates them (Buchner et al., 2019). Climate finance is particularly useful for climate actions in developing countries with limited financial resources and technical know-how. In catalyzing additional investments from the private sector, public funding can help de-risk investments. Climate finance thus not only expedites but expands governmental actions for energy transition (Bhandary et al., 2021).

Given the significance of climate finance, PA Article 3 requires all parties to make ambitious efforts towards "making finance flows consistent with a pathway towards low greenhouse gas emissions and climate-resilient development." Progress towards this objective is also part of the global stocktake. It can potentially send a strong signal to all relevant actors, including the private sector, to reassess and redirect investments.

During negotiations of the PA, policymakers had extensively debated whether the PA should anchor and continue the political commitment made in Copenhagen to mobilize US $100 billion per year by 2020 or even specify higher amounts. However, the PA reaffirmed the continuing existing obligations under the UNFCCC, not resetting quantified obligations or setting up a reference to the financial commitment.

Accordingly, the Green Climate Fund (GCF) and the Global Environmental Facility (GEF) were chosen to become the operating entities for climate finance. The GCF is committed to building the resilience of developing countries by supporting nations' climate adaptation efforts. It aims to deliver a one-on-one balance between adaptation and mitigation in its portfolio, ensuring that at least 50% of adaptation funding goes to particularly vulnerable countries, including least developed countries (LDCs), small island developing states (SIDS), and African states. A grant of up to US $3 million is in place to formulate national adaptation plans (NAPs) and/or other adaptation planning processes. Since mid-2018, 23 countries have requested and received technical assistance to prepare adaptation planning proposals (GCF, 2022).

The Adaptation Fund is another climate financing mechanism serving the new agreement (Articles 9 and 10). The fund finances projects and programs that help vulnerable communities in developing countries adapt to climate change. Initiatives are based on the country's needs and priorities. All developing country parties to the Kyoto Protocol are eligible to nominate an entity for accreditation. Once an entity passes the fund's rigorous accreditation review, it may apply for project funding. Through direct access, accredited National Implementing Entities (NIEs) can directly access financing and manage all aspects of climate adaptation and resilience projects, from design through implementation and monitoring (Adaptation Fund, 2019). Since 2010, the Adaptation Fund has committed over US $1 billion for climate change adaptation and resilience projects and programs, including 150 concrete, localized projects in the most vulnerable communities of developing countries, with over 38 million total beneficiaries (Adaptation Fund, 2023).

REDD+ Framework for a PA Implementation Mechanism

Countries established the REDD+ framework as part of the Paris Agreement. REDD stands for Reducing Emissions from Deforestation and Forest Degradation. The plus stands for the role of conservation, sustainable management of forests, and enhancement of forest carbon stocks. Under the REDD+ framework, developing countries can conserve forests and invest in low-carbon paths to sustainable development while reducing deforestation.

REDD+ promotes results-based payments and climate-equitable development (Christen et al., 2020), and many developed countries such as Norway, Germany, and UK have funded REDD+ activities. GCF and other financial mechanisms also provide financial support. Technologies like remote sensing play a crucial role in monitoring deforestation and forest degradation, while addressing climate externalities, the biggest market failure in human history (Stern, 2006).

In recognition of economic incentives, some developing countries, such as Brazil, Indonesia, and Costa Rica, have incorporated REDD+ into their climate action plans. Costa Rica received the first emission reductions payment of US $16.4 million for reducing 3.28 million tons of carbon emissions during 2018 and 2019 (World Bank, 2022).

As of 2022, REDD+ activities implemented by developing countries cover a forest area of approximately 1.35 billion hectares, which accounts for about 62% of forest area in developing countries and about 75% of global deforestation. The UN Climate Change secretariat has been undertaking REDD+ technical assessments for 10 years. In total, 60 developing countries have reported REDD+ activities. As a result of REDD+ activities, 16 of these countries reported a reduction of almost 11 billion tons of carbon dioxide, almost twice the amount of net greenhouse gas emissions from the United States in 2021, and are now eligible to seek results-based finance (UNFCCC, 2022).

Despite its contribution to carbon reduction and forest conservation, REDD+ faces several challenges. First, the idea behind results-based payments is to reward developing countries for verified emissions reductions. However, the benefits of REDD+ activities might not reach to local community members who manage forests. This means that payment schemes often favor national governments or large organizations, potentially sidelining indigenous and local communities. Prior research suggests that results-based payments must be accompanied by strong safeguards to ensure that the rights and livelihoods of local communities are not negatively impacted (Jagger et al., 2014).

Secondly, the process of measuring, reporting, and verifying carbon credits presents challenges to REDD+ results-based activities. The process of measurement, reporting, and verification (MRV) involves multi-steps to *measure* GHG emissions reduction by specific mitigation activities, such as reducing emissions from deforestation and forest degradation, over a period of time and to *report* these findings to an accredited third party. The third party then *verifies* the report so that carbon credits can be issued by certified credits. A recent study indicated that a dedicated verification procedure is highly needed for developing countries to access the results-based financing (Christen et al., 2020).

Also, there are concerns about the long-term permanence of emissions reductions as well as carbon leakage in which deforestation is just shifted to other regions (Streck, 2021). Emissions leakage refers to the spillover effects of a mitigation measure. More specifically, leakage is the decrease or increase of GHG reductions and removals outside of a project or program boundary (Atmadja & Verchot, 2012). As such, leakage renders mitigation measures ineffective by shifting emissions to another place rather than abating them. The risk of emissions leakage remains one of the most salient quality issues besetting REDD+ implementation, in particular with project-level activities (Seymour, 2020).

Beyond Paris: Collaborative Governance Critical to Decarbonization

Since its inception in 1992, the UNFCCC has served as the global anchor for climate policy, setting a broad strategy for climate action and promoting capacity building for developing countries. The 1997 Kyoto Protocol has substantiated global climate mitigation actions by mandating that 38 developed nations reduce greenhouse gases on average 5.2% of the 1990 baseline emissions in the 2008–2012 period. More recently, the 2015 PA committed all governments collectively to, among other goals, reduce emissions enough to limit global temperature rise well below 2 degrees Celsius and aim for 1.5 degrees Celsius above preindustrial levels. The triumphant trajectory of global climate regimes underpins the urgency of climate actions and countries' political will to achieve carbon reductions to avoid climate risks and disasters.

In correspondence to the solidified climate policy architecture, the transformation of economies is required for decarbonization and necessitates the engagement of many relevant actors. In this vein, the PA needs to provide a strong policy signal that ending the long-term business case for fossil fuels is a decisive step, and meaningful climate actions should be ensured by different levels of government, from civil society and business. It is known that climate action by nonstate and subnational actors has been complementary to national policies (Andonova et al., 2017; Roger et al., 2017), and the mitigation potential of such action has been significant. Although some countries roll back climate policies, businesses and nonstate actors increasingly commit themselves to deeper and more ambitious climate actions (Kuramochi et al., 2020).

In particular, environmentally progressive businesses have also played a pivotal role in carbon-reducing projects, leading carbon markets, and making substantive private investments (Kim, 2022; Kim & Darnall, 2016). The leadership on climate actions by multinational global businesses was confirmed at the 26th COP held in Glasgow in 2022. For instance, more than 5,000 businesses pledged to meet net-zero carbon targets by 2050. In addition, some 450 banks, insurers, and investors that collectively represent US $130 trillion in assets and 40% of the world's private capital has committed to make their portfolios climate-neutral during the same period. A group of major automakers pledged to stop selling gasoline-powered vehicles by 2035. Those commitments spotlight businesses' leadership triggered by the 2015 PA and unchanging commitments despite the external business environmental change, such as the global pandemic crisis (Yanosek & Victor, 2022).

Likewise, collaborative climate governance can expand private investments in climate actions, speed up climate adaptation in developing countries, and scale up GHG emissions reductions by technologies, supporting countries' achievements of nationally determined contributions (Kuramochi, 2020). It is thus clear that mobilizing sufficient resources and creativity at all levels of social organization can facilitate a global transition to a low-carbon future. The gloomy forecast of the PA's temperate goal achievement (Mooney, 2023) could be overcome in this way.

References

Adaptation Fund. (2019). *Climate adaptation finance: Direct access.* https://www.adaptation-fund.org/wp-content/uploads/2018/04/Direct-Access-English-May-2019-WEB.pdf

Adaptation Fund. (2023). *About the Adaption Fund.* https://www.adaptation-fund.org/about/

Andonova, L. B., Hale, T. N., & Roger, C. B. (2017). National policy and transnational governance of climate change: Substitutes or complements? *International Studies Quarterly, 61*(2), 253–268. https://doi.org/10.1093/isq/sqx014

Atmadja, S., & Verchot, L. (2012). A review of the state of research, policies and strategies in addressing leakage from reducing emissions from deforestation and forest degradation (REDD+). *Mitigation and Adaptation Strategies for Global Change, 17*(3), 311–336. https://doi.org/10.1007/s11027-011-9328-4

Bodansky, D., Brunnée, J., & Hey, E. (2007). International environmental law: Mapping the field. In D. Bodansky, J. Brunnée, & E. Hey (Eds.), *The Oxford handbook of international environmental law.* Oxford University Press, pp. 1–26.

Bodle, R., Donat, L., & Duwe, M. (2016). The Paris Agreement analysis, assessment and outlook. *Carbon & Climate Law Review, 10*(1), 5–22. https://cclr.lexxion.eu/article/CCLR/2016/1/4

Bhandary, R. R., Gallagher. K. S., & Fang. Z. (2021). Climate finance policy in practice: A review of the evidence. *Climate Policy, 21*(4), 529–545. https://doi.org/10.1080/14693062.2020.1871313

Brunnée, J. (2012). Treaty amendments. In D. B. Hollis (Ed.), *The Oxford guide to treaties.* Oxford University Press, pp. 347–366.

Brunnée, J. & Streck, C. (2013). The UNFCCC as a negotiation forum: Towards common but more differentiated responsibilities. *Climate Policy, 13*(5), 589–607. https://doi.org/10.1080/14693062.2013.822661

Buchner, B., Clark, A., Falconer, A., Macquarie, R., Meattle, C., Tolentino, R., & Cooper, W. (2019). *Global landscape of climate finance 2019.* Climate Policy Initiative. https://climatepolicyinitiative.org/wp-content/uploads/2019/11/2019-Global-Landscape-of-Climate-Finance.pdf

Christen, D. R., Espinosa, M.G., Reumann, A., puri, J. Results-based payments for REDD+ under the Green Climate Fund: Lessons learned on social, environmental, and governance safeguards. *Forests, 11*(12):1350. https://doi.org/10.3390/f11121350

CO2.earth (2023). *Latest daily CO2.* https://www.co2.earth/daily-co2

Falkner, R. (2016). The Paris Agreement and the new logic of international climate politics. *International Affairs. 92*(5), 1107–1125. https://doi.org/10.1111/1468-2346.12708

Gehring, T. (2007). Treaty-making and treaty evolution. In D. Bodansky, J. Brunnée, & E. Hey (Eds.), *The Oxford handbook of international environmental law.* Oxford University Press, pp. 467–497.

Green, J. F. (2021). Does carbon pricing reduce emissions? A review of ex-post analyses. *Environmental Research Letters, 16*(4), 43004. https://doi.org/10.1088/1748-9326/abdae9

Green Climate Fund (GCF) (2022). Catalyzing action and finance for country adaptation priorities. GCF In Brief: Adaptation Planning. https://www.greenclimate.fund/sites/default/files/document/20221102-gcfin-brief-adaptation-planning.pdf

IPCC (1996). *Climate change 1995: The science of climate change.* Cambridge University Press.

Jagger, P., Brockhaus, M., Duchelle, A. E., Gebara, M. F., Lawlor, K., Resosudarmo, I.A.P., & Sunderlin, W. D. (2014). Multi-level policy dialogues, processes, and actions: Challenges and opportunities for national REDD+ safeguards measurement, reporting, and verification (MRV). *Forests, 5*(9), 2136–2162. https://doi.org/10.3390/f5092136

Jordan, A. J., Huitema, D., Hildén, M., van Asselt, H., Rayner, T. J., Schoenefeld, J. J., Tosun, J., Forster, J., & Boasson, E. L. (2015). Emergence of polycentric climate governance and its future prospects. *Nature Climate Change, 5,* 977–982. https://doi.org/10.1038/nclimate2725

Kim, R. E., & Mackey, B. (2014). International environmental law as a complex adaptive system. *International Environmental Agreements, 14*(1), 5–24. https://doi.org/10.1007/s10784-013-9225-2

Kim, Y. (2022). Integrated market and nonmarket strategies: Empirical evidence from the S&P 500 firms' climate strategies. *Business and Politics, 24,* 1–22. https://doi.org/10.1017/bap.2021.18

Kim, Y. (2023). *Carbon markets: Principles and current practices.* EBSCO Pathways to Research.

Kim, Y. & Darnall, N. (2016). Business as a collaborative partner: Understanding firms' sociopolitical support for policy formation. *Public Administration Review, 76*(2), 326–337. https://doi.org/10.1111/puar.12463

Kuramochi, T., Roelfsema, M., Hsu, A., Lui, S., Weinfurter, A., Chan, S., Hale, T., Clapper, A., Chang, A., & Höhne, N. (2020). Beyond national climate action: The impact of region, city, and business commitments on global greenhouse gas emissions. *Climate Policy, 20*(3), 275–291. https://doi.org/10.1080/14693062.2020.1740150

Mooney, A. (2023, October 26). UN science body head fears lower chance of keeping global warming to 1.5 °C. *Financial Times.* https://www.ft.com/content/2937f8a2-e553-4a60-aadf-553cd7f2d4e3

Roger, C., Hale, T., & Andonova, L. (2017). The comparative politics of transnational climate governance. *International Interactions, 43*(1), 1–25. https://doi.org/10.1080/03050629.2017.1252248

Selin, H., & VanDeveer, S. D. (2022). Global climate change governance: Can the promise of Paris be realized? In N. J. Vig, M. E. Kraft, & B. G. Rabe (Eds.), *Environmental policy: New directions for the twenty-first century* (pp. 275–299) CQ Press.

Seymour, F. (2020, May 5). INSIDER: 4 reasons why a jurisdictional approach for REDD+ crediting is superior to a project-based approach. World Resources Institute. https://www.wri.org/insights/insider-4-reasons-why-jurisdictional-approach-redd-crediting-superior-project-based

Stern, N. (2006). *The economics of climate change: The Stern review.* https://webarchive.nationalarchives.gov.uk/ukgwa/20100407172811/https://www.hm-treasury.gov.uk/stern_review_report.htm

Streck, C. (2021) REDD+ and leakage: Debunking myths and promoting integrated solutions. *Climate Policy, 21*(6), 843–852. https://doi.org/10.1080/14693062.2021.1920363

UNFCCC. (2022). *What is REDD+?* https://unfccc.int/topics/land-use/workstreams/redd/what-is-redd#The-REDD-success-story

Victor, D. G., (2011). *Global warming gridlock: Creating more effective strategies for protecting the planet,* Cambridge University Press.

World Bank. (2022). *Costa Rica receives first emission reductions payment form Forest Carbon Partnership Facility.* https://www.worldbank.org/en/news/press-release/2022/08/16/-costa-rica-receives-first-emission-reductions-payment-from-forest-carbon-partnership-facility

Yanosek, K & Victor, D. G. (2022). How big business is taking the lead on climate change: Harnessing private capital for the public good. *Foreign Affairs,* https://www.foreignaffairs.com/articles/world/2022-02-03/how-big-business-taking-lead-climate-change

DISCUSSION QUESTIONS

1. An important yardstick of the Paris Agreement's success is its ability to increase ambition over time via the cycles of NDCs. What are NDCs? Do you believe that NDC cycles are strongly designed to ensure the Paris Agreement is accomplished? Why? If not, how can NDCs be improved?

2. One of the unique peculiarities of the Paris Agreement is climate finance. Why is climate finance considered to be a key element? What are the appropriate roles for the Green Climate Fund and the Adaptation Fund?

3. The PA recognizes the role of forests in mitigating and adapting to climate change and adopts the REDD+ framework to reduce emissions from deforestation and forest degradation, enhance the conservation and management of forests, and increase forest carbon stocks in developing countries. What is the current state of REDD+ framework? Do you believe that the results-based payments would promote climate-equitable development in the Global South? If not, what are the mechanisms that could support REDD+ framework function more appropriately?

Energy Transition

From Fossil Fuels to Renewables for a Net-Zero,
Climate Resilient World

By Younsung Kim

Introduction

The energy transition is currently a major focus worldwide. Energy transition refers to the shift from one dominant energy resource or a set of energy sources to another. The transition from whale oil to kerosene and wood to coal during the 19th century exemplifies a previous energy transition. The present energy transition is marked by moving away from carbon-intensive fossil fuels toward renewables with low carbon intensity (Carley et al., 2018).

Renewables have been growing since the 2000s. As of 2022, renewables drive 29% of the world's energy (IEA, 2022). A notable achievement has been observed in electricity generation, a sector known for its high carbon intensity. In 2022, carbon-free energy sources, including nuclear power, hydropower, and various renewables, will account for 38% of the world's electricity. As a landmark milestone, observed solar photovoltaic's (PV) share of global electricity generation surprisingly rose from 3.6% to 4.5% from 2021 to 2022 (Dreves, 2023). Renewable technology breakthroughs, affordable prices, and institutional support combined have led to the achievement and offer a positive sign for energy transition (Haegel & Kurtz, 2023).

Energy policies in favor of renewables shape the direction and speed of the energy transition. In the U.S. context, some states like California have been active in divesting fossil fuels through energy policies, including Renewable Portfolio Standard (RPS). RPS is a mandatory policy in which a specific percentage of total state electricity sales, or generation, is sourced from renewable energy. In other institutional contexts, another favored renewable policy instrument is the Feed-in Tariff (FIT), which guarantees a price for renewable energy producers for each unit of energy produced and fed into the grid. The explosive growth of renewable electricity in Germany since 2000 is attributed to the FIT that offered strong incentives to energy suppliers (Heinberg & Fridley, 2016; Karazin, 2020).

This chapter discusses key policy instruments for energy transition, focusing on RPSs and FITs. It also outlines the latest renewable policy initiatives in the European Union, the United States, and

China, the world's top greenhouse gases (GHG) emitters. The justice implication of energy transition is also reviewed, as energy transition may inevitably produce and perpetuate inequality associated with energy production and supply. In doing so, this chapter illuminates the role of government in scaling up and accelerating energy transition and decarbonization. Governments can work with the private sector for technology development and infrastructure changes while protecting marginalized communities from bearing disproportionate burdens or costs that accompany the energy transition.

Renewable Policy Tools

Renewable energy sources offer a multifaceted solution to various challenges posed by the current fossil fuel-based energy system. First, renewables can reduce a myriad of environmental pollution in air, water, and soil, which affects human and ecological health seriously. For instance, coal mining alters the landscape of forested and mountainous areas, pollutes nearby streams and groundwater reservoirs, and causes peoples' health issues like asthma and lung cancer. Furthermore, abandoned mine lands require ecological remediation and ecosystem restoration that demand public budgeting (Karsyn & Kim, 2023). Energy transition away from fossil fuels thus can lower environmental pollution in the lifecycle of use of fossil fuels. Second, the transition to renewables drives technological innovations for sustainable development and stimulates economic growth through new markets and job creation (Rokeach & Schatz, 2012; Ostergaard et al., 2020). Sustainable development in developing countries could only be possible when development strategies include renewable energy technologies and renewables as part of their energy system (Cantarero, 2020; Gu et al., 2018; González et al., 2017; Ouedraogo, 2017; Vivaili et. al, 2017; Wesseh & Lin, 2015).

However, transitioning to renewable energy sources presents significant challenges and constraints. The capital-intensive nature of changing energy systems, coupled with the need for coordination among various stakeholders, can impede progress. Additionally, the long-term nature of the benefits from renewable investments poses challenges for investors and project developers who may seek more immediate returns (Kim & Oh, 2017). To address these constraints, policies aimed at stimulating both demand for renewable electricity and supply of renewable energy are crucial. These policies can take various forms, including RPS, FITs, tax incentives and subsidies, carbon pricing, or research and development funding. Here, we focus on RPS and FITs.

Renewable Portfolio Standard (RPS)

RPS is a performance-based command-and-control type of policy. It requires a certain percentage of electricity sales or production by utilities to come from renewable sources. For instance, in 2020, Virginia enacted a mandatory 100% renewable target for utilities, requiring utilities to procure a certain amount of generation from solar and onshore wind sources located in the state. RPS thus incites renewable market developments by creating stable, increasing demand for renewables (Komor & Brazilian, 2005). It also encourages investment with a clear target for renewable energy production.

However, RPS does not guarantee a fixed price for renewable energy, which can lead to market price volatility and uncertainty for investors.

Although national RPS has been proposed, no federal RPS or similar policy exists in the United States. Instead, states have enacted their own RPS to supply renewable-based electricity while filling the void of federal inaction in climate mitigation (Rabe, 2022). State programs have been varied, as some states focus the RPS requirements on large investor-owned utilities, while others apply the standards to all utilities operating in the state. Likewise, state-driven RPS has shown differences in program structure, enforcement mechanisms, size, applications, and so forth, which could relate to performance (Delmas & Montes-Sancho, 2011; Carley et al., 2018).

As of November 2022, 36 states and the District of Columbia had established RPS or renewable energy goals that are voluntary targets. In 12 of those states and the District of Columbia, the requirement is for 100% clean electricity by 2050 or earlier. Figure 1.3.1 introduces a summary map and details on state RPS programs available in the Database of State Incentives for Renewables and Efficiency (DSIRE) (US EIA, 2024).

RPS policies are known to be effective (Fischlein & Smith, 2013; Wiser et al., 2011), as evidenced by actual renewable energy growth outpaced by RPS needs in Texas and the Northeast, Mid-Atlantic, and Western regions (Barbose, 2023). Prior research also found that more stringent RPS has led to more renewable energy development, thereby indicating that states without RPS can consider

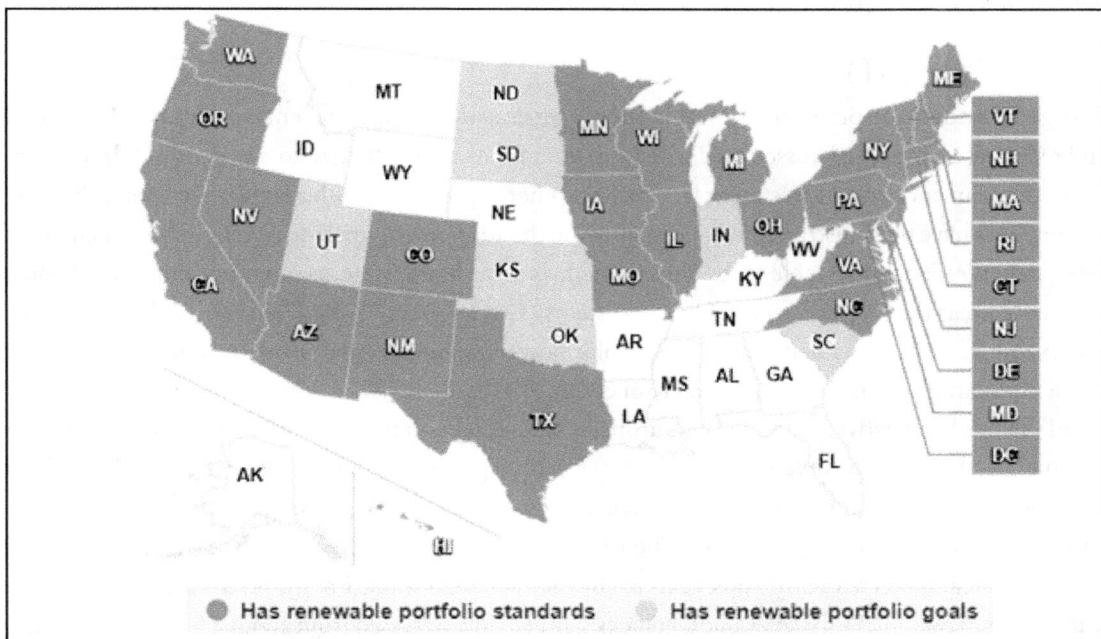

Figure 1.3.1 States adopted RPS and Renewable Portfolio Goals.
Source: US Energy Information Agency (EIA), https://www.eia.gov/energyexplained/renewable-sources/portfolio-standards.php, 2022.

ambitious goals, as well as mechanisms encouraging in-state renewable growth. Most RPSs allow utilities to buy credits from renewable energy certificate markets to satisfy their RPS requirements. Thus, banning renewable energy certificate trades can be more likely to boost in-state renewable electricity production (Carley et al., 2018).

Despite effectiveness, RPS-dependent renewable energy development can cause uneven state performance, and the spillover effect of RPS has been minimal (Rabe, 2022). This means that some states like North Dakota have adopted renewable energy goals as voluntary targets, while many Southeastern states have not yet adopted RPS in part owing to their low electricity cost and limited options of affordable renewables such as wind. In those states, the growth of renewables is slow, and fossil fuel-based electricity supply is still dominant.

Moreover, in many cases, RPS could drive higher electricity prices. There are system costs of increased renewable energy use, including transmission upgrades, integration costs, and planning and operation-related transaction costs. For instance, when renewable-based utilities are built in remote or distant areas, they may require additional transmission lines, which is an extra cost to the base cost for the electricity system. Concerning integration, generation from wind and solar facilities is variable and uncertain, requiring additional planning and perhaps operational changes. Accordingly, to the extent that renewables increase the number of small, distributed generation installations, managing the interconnection and integration of many small-scale producers can increase administrative and operational burdens (Leon, 2012).

Feed-in Tariff (FIT)

FITs guarantee a fixed price for renewable energy generated, providing revenue certainty for investors and reducing financial risks associated with long-term projects to both investors and project developers. Thus, they can encourage investment by providing energy producers with a stable, predictable income stream. FITs have been instrumental in encouraging the adoption of small-scale renewable installations using decentralized renewable technologies, such as solar home systems or community microgrids, in remote areas that are not served by grid systems (Goldemberg, 2006; Heinberg & Fridley, 2016). For example, an FIT enabled a small-scale solar power generation that used rooftops of public and residential buildings in Gujarat, India (Kim & Oh, 2017).

FITs have been actively used in the European Union, including Germany and Spain. In 1991, the German Feed-In Law (*Stromeinspeisungsgesetz*) required power utilities to buy electricity generated by renewable energy providers, with wind and solar electricity producers guaranteed 90% of the retail rate and 75% for biomass and hydropower. The FIT rate needed to be increased to encourage more solar development and was reformed and expanded by the 2000 Renewable Energy Sources Act. As a unique feature of the law, the FIT for solar photovoltaics significantly increased from about 8.5 to 51-euro cents/kilowatt hour (kWh) and guaranteed rates to renewable electricity producers for the first 20 years of a facility's operation. Owing to such FITs, Germany exceeded its federal targets for renewable electricity by 20% by 2020 and increased the 2030 target from 50% to 60% of electricity consumption (Karazin, 2020).

The Spanish government also instituted an FIT in 1996 and required utilities to purchase electricity generated by renewables at premium rates. Power companies started building wind farms, and Spain's renewable subsidies led to a nearly 40-fold increase in wind capacity to 16.7 GW in 2008. In 2004, the Spanish government also instituted a generous FIT of 46-euro cents/kilowatt hour (kWh) for solar electricity. Investors responded to long-term profits, and the rates of solar installation soared. In 1998, Spain installed more than 2.5 GW of solar PV capacity, exceeding the 2008 national target of 400 MW in installed solar capacity (Heinberg & Fridley, 2016).

Latest Policy Development for Energy Transition in United States, European Union, and China

The recent progress in renewable energy growth worldwide has been remarkable. The year 2022 was a record year for renewable electricity capacity additions, with annual capacity additions amounting to about 340 GW. Countries highly responsible for global GHG emissions announced policies to facilitate energy transition. Key policies are REPowerEU in the European Union, Inflation Reduction Act (IRA) in the United States, and China's 14th Five-Year Plan for Renewable Energy. These policies are expected to support scaling up and accelerating renewable electricity deployment and ensure energy security.

REPowerEU is the European Union's timely, strategic response to hardships and global energy market disruptions caused by Russia's invasion of Ukraine. The central aim of REPowerEU is twofold. First, it intends to end Europe's dependence on Russian fossil fuels as quickly as possible, in principle by 2027, with two thirds of cut in Russian gas consumption by the end of 2022. Second, it desires to secure long-term sustainability, cost-effectiveness, and energy supply to the EU energy system by demand reduction, diversification of suppliers for fossil fuel imports, and accelerating the clean energy transition. Given the objectives, the European Union targeted 320 GW of solar PV capacity installed by 2025, rising to 600 GW by 2030. New policies and targets proposed in the REpowerEU are expected to be important drivers of renewable energy investments in the coming years (EU, 2024; IEA 2023).

The United States enacted the Inflation Reduction Act (IRA) in 2022. This act signifies a substantial commitment to the energy transition and climate change mitigation. By earmarking approximately $370 billion for clean energy and climate-related measures, the law draws the largest investment in combating climate change in U.S. history. It aims to address the urgent need for energy transition while also tackling the pressing issue of inflation, exacerbated by a global energy crisis. The law includes a combination of grants, loans, tax provisions and other incentives to accelerate the deployment of clean energy, clean vehicles, clean buildings, and clean manufacturing. Diverse incentives and grants aim to reduce renewable energy costs for organizations, including businesses, nonprofits, educational institutions, and state, local, and tribal organizations and thereby scale up renewable energy expansion (US EIA, 2024).

China's 14th Five-Year Plan for Renewable Energy for the period of 2021–2025 is also noteworthy, as it includes the country's GHG emissions peak by 2030 and energy transition as a crucial pillar to achieve carbon neutrality by 2060. This plan is particularly significant as it could support the country's recovery from the economic downturn caused by the COVID-19 crisis, while affirming its commitment to sustainable economic growth by reducing investments in coal-based power generation. In the context of peaking GHG emissions by 2030, the country indeed added 160 GW of renewable electricity capacity in 2022, which is almost half of all global deployment, signifying the country's shifting perspective from old coal-intensive development paradigm to long-term prosperity (Hepburn et al., 2021; IEA, 2023).

The European Union, the United States, and China are the top three carbon emitters, accounting for up to 50% of global greenhouse gas emissions. Their actions are deterministic for the ambitious target of the Paris Agreement, limiting global temperature increases to well below 2 degrees Celsius, while pursuing efforts to limit the increase to 1.5 degrees Celsius. Accordingly, the latest policy development in those three countries reflects an understanding of the critical role that governments play in promoting energy transition and achieving carbon neutrality in alignment with the Paris climate agreement (Skjærseth et al., 2021; Zhu, 2021).

Justice Implication of Energy Transition

Energy transition must be just, as it will potentially affect socioeconomically marginalized groups and communities, particularly those whose livelihoods are deeply entwined with traditional energy sectors like coal and oil. The decline of these industries, while environmentally beneficial, can lead to job losses, economic downturns, and social dislocation for the communities involved. Recognizing these challenges, various governments and international bodies have started to implement policies and initiatives that aim to address the socioeconomic impacts of the energy transition. The EU's Green New Deal (EC, 2019) and the U.S. Green New Deal proposals, along with legislative acts such as the U.S. Bipartisan Infrastructure Law and the IRA, are prime examples of this approach. These initiatives seek to ensure that the move towards a sustainable energy future is inclusive and equitable, offering support for those most affected by the transition (Fiorino, 2022; Karsyn & Kim, 2023).

The concept of a "just transition" encompasses a range of strategies, including retraining programs for workers in traditional energy sectors, investments in developing new industries in regions dependent on fossil fuels, and social support mechanisms to assist communities in adapting to economic changes. These measures are designed to ensure that the benefits of the energy transition are shared widely and that the burdens are not disproportionately borne by the most vulnerable populations (Carley & Konisky, 2020; Karsyn and Kim, 2023).

The Biden administration particularly put a high emphasis on just transition, integrating energy transition with environmental justice. Breaking away from the Trump administration's inaction on climate policy, the Biden administration has declared climate change not only a national threat but also an indispensable opportunity to rejuvenate the U.S. energy and manufacturing sectors, potentially creating millions of high-quality, well-paying jobs. It set the goal of a 100% clean energy economy and net-zero emissions no later than 2050, and the U.S. Department of Energy has launched the Office of State and Community Energy Programs (SCEP) to implement nearly $16 billion programs from the Infrastructure Law, IRA, and annual appropriations. Programs led by SCEP are designed to help communities nationwide significantly accelerate the deployment of clean, reliable energy technologies and ensure that households and businesses see lower energy bills and other benefits. There are several grants in the SCEP, mostly related to technical and financial assistance, training and assessment for energy transition, as well as fostering capacities in states, tribes, territories, local governments, nonprofits and schools to catalyze local economic development (US DOE, 2023a).

Some SCEP grants contribute to the Biden administration's Justice40 Initiative, which aims to address historical and systematic environmental injustices with the goal of directing 40% of the overall benefits from certain federal investments to historically disadvantaged communities that are marginalized, underserved, and overburdened by pollution. For instance, the U.S. Department of Energy (DOE) launched the $18 million Local Energy Action Program (LEAP) to assist to low-income, energy-burdened communities that are also experiencing either direct environmental justice impacts or direct economic impacts from a shift away from historical reliance on fossil fuels. Under Communities LEAP, selected communities receive technical assistance to help them develop clean energy planning and economic development visioning. In 2022, 24 communities that are historically reputed EJ communities or coal towns were selected to get the needed support to create community-level clean energy targets and plan toward energy transition (US DOE, 2023b)

Those policy initiatives ensure energy transition does not exacerbate preexisting energy insecurity and inequality associated with negative externalities of pollution. They also encourage careful planning and multi-stakeholder engagement by providing the needed information for the positive benefits side of the energy transition so that they can meaningfully engage in the decision-making process of selecting brownfields or abandoned coal mining lands for ecological remediation and sustainable economic development (Carley & Konisky, 2020; Karsyn & Kim, 2023; Markkanen & Anger-Kraavi, 2019).

Conclusion

The shift toward renewables and away from fossil fuels is one of the biggest changes in global energy supply. Renewable electricity deployment worldwide has been increasing, and some renewable policies, including RPSs and FITs, have proven successful, creating demand for renewable electricity and supporting more reliable, long-term supply with price adjustments. However, there is still a significant gap between today's action on net-zero pledges and limiting 1.5 degrees Celsius of warming, and countries should be strategic toward energy transition. A sectoral approach focusing on carbon-intensive sectors such as building, industrial manufacturing and mining, and transportation could usher in a more assured, cost-effective pathway to climate mitigation and transition to a clean energy economy (Harvey et al., 2018).

Also, partnerships and collaboration among governments, the private sector, and civil society are key to energy transition insofar as industrial opposition to climate mitigation coexists with social support for renewables (Kim & Darnall, 2016). It is evident that clean electrification and decarbonization come at a cost, and collaborative governance can make it easier to secure broad-based support for energy transition. The notion of just transition is critical, and governments should continue to work with local communities and historically disadvantaged groups so that the transition's benefits are widely shared and inclusive. The latest policy initiatives in the world's top GHG emitters give hope for accelerating energy transition and building resilient communities capable of thriving in a net-zero, climate-resilient world.

References

Cantarero, M. M. V. (2020) Of renewable energy, energy democracy, and sustainable development: A roadmap to accelerate the energy transition in developing countries, *Energy Research & Social Science, 70*, 101716. https://doi.org/10.1016/j.erss.2020.101716.

Carley, S., Davies, L. L., Spence, D. B., & Zirogiannis, N. (2018). Empirical evaluation of the stringency and design of renewable portfolio standards. *Nature Energy, 3*, 754–763. https://doi.org/10.1038/s41560-018-0202-4

Carley, S., & Konisky, D. M. (2020). The justice and equity implications of the clean energy transition. *Nature Energy, 5*, 569–577. https://doi.org/10.1038/s41560-020-0641-6

Delmas, M. A., & Montes-Sancho, M. J. (2011). U.S. state policies for renewable energy: Context and effectiveness. *Energy Policy, 39*(5), 2273–2288. https://doi.org/10.1016/j.enpol.2011.01.034

Dreves, H. (2023). *At a glance: How renewable energy is transforming the global electricity supply*. National Renewable Energy Laboratory. https://www.nrel.gov/news/program/2023/how-renewable-energy-is-transforming-the-global-electricity-supply.html

European Commission. (2019). *The European Green Deal.* https://commission.europa.eu/strategy-and-policy/priorities-2019-2024/european-green-deal_en

European Union (EU). (2024). *REPowerEU: Affordable, secure and sustainable energy for Europe.* https://commission.europa.eu/strategy-and-policy/priorities-2019-2024/european-green-deal/repowereu-affordable-secure-and-sustainable-energy-europe_en

Fiorino, D. J. (2022). Creating the green economy: government, business, and a sustainable future. In *Environmental Policy: New Directions for the Twenty-First Century,* In *Environmental Policy: New Directions for the Twenty-First Century,* In N. J. Vig, M. E. Kraft, & B. G. Rabe (Eds.), *Environmental policy: New directions for the twenty-first century* (pp. 323–344). CQ Press.

Fischlein, M. & Smith, T. M. (2013). Revisiting renewable portfolio standard effectiveness: Policy design and outcome specification matter. *Policy Sciences, 46*, 277–310. https://doi.org/10.1007/s11077-013-9175-0

Goldemberg, J. (2006). The case for renewable energies. In D. Assmann, U. Laumanns, and D. Uh. (Eds.), *Renewable energy: A global review of technologies, policies and markets*. Earthscan, pp. 3–14.

González, M. O. A., Gonçalves, J. S., & Vasconcelos, R. M. (2017). Sustainable development: Case study in the implementation of renewable energy in Brazil. *Journal of Cleaner Production, 142*, 461–475. https://doi.org/10.1016/j.jclepro.2016.10.052

Gu, J., Renwick, N., & Xue, L. (2018). The BRICS and Africa's search for green growth, clean energy and sustainable development. *Energy Policy, 120*, 675–683. https://doi.org/10.1016/j.enpol.2018.05.028

Haegel, N. M., & Kurtz, S.R. (2023). Global progress toward renewable electricity: Tracking the role of solar, *IEEE Journal of Photovoltaics, 13*(6): 768–776, doi: 10.1109/JPHOTOV.2023.3309922.

Harvey, H., Orvis, R., & Rissman, J. (2018). *Designing climate solutions: A policy guide for low-carbon energy*. Island Press.

Heinberg, R., & Fridley, D. (2016). *Our renewable future: Laying the path for one hundred percent clean energy*. Island Press.

Hepburn, C., Qi, Y., Stern, N. Ward, B., Xie, C., & Zenghelis, D. (2021). Towards carbon neutrality and China's 14th Five-Year Plan: Clean energy transition, sustainable urban development, and investment priorities, *Environmental Science and Ecotechnology, 8*, 100130. https://doi.org/10.1016/j.ese.2021.100130

International Energy Agency. (2022). *Renewables 2022: Analysis and forecast to 2027*. https://iea.blob.core.windows.net/assets/ada7af90-e280-46c4-a577-df2e4fb44254/Renewables2022.pdf

International Energy Agency. (2023). *Renewables 2023: Analysis and forecast to 2028*. https://iea.blob.core.windows.net/assets/96d66a8b-d502-476b-ba94-54ffda84cf72/Renewables_2023.pdf

Karazin, R. (2020). Household costs and resistance to Germany's energy transition. *Review of Policy Research, 37*(3), 313–341. https://doi.org/10.1111/ropr.12371

Karsyn, K., & Kim, Y. (2023). Abandoned mine land program: Examining public participation in decision-making. *Review of Policy Research*, 1–29. https://doi.org/10.1111/ropr.12569

Kim, Y., & Darnall, N. (2016). Business as a collaborative partner: Understanding firms' sociopolitical support for policy formation. *Public Administration Review, 76*(2), 326–337. https://doi.org/10.1111/puar.12463

Kim, Y., & Oh, K. (2017). Public and private partnerships for enhanced energy access in developing countries: A case of the solar rooftop project in Gujarat, India. In J. Leitão, E. de Morais Sarmento, & J. Aleluia. (Eds.), *Handbook on PPPs in developing and emerging economies*. (pp. 537-557). Emerald Group Publishing.

Komor P., & Brazilian, M. (2005). Renewable energy policy goals, programs, and technologies. *Energy policy, 33*(14), 1873–1881. https://doi.org/10.1016/j.enpol.2004.03.003

Leon, W. (2012). *Evaluating the benefits and costs of a renewable portfolio standard: A guide for state RPS programs*. Clean Energy States Alliance. https://www.cesa.org/wp-content/uploads/CESA-RPS-evaluation-report-final-5-22-12.pdf

Markkanen S., & Anger-Kraavi, A. (2019). Social impacts of climate change mitigation policies and their implications for inequality. *Climate Policy, 19*(7), 827–844. https://doi.org/10.1080/14693062.2019.1596873

Østergaard, P. A., Duic, N., Noorollahi, Y., & Kalogirou, S. (2020). Latest progress in sustainable development using renewable energy technology, *Renewable Energy, 162,* 1554–1562. https://doi.org/10.1016/j.renene.2020.09.124

Ouedraogo, N. S. (2017). Africa energy future: Alternative scenarios and their implications for sustainable development strategies. *Energy Policy, 106,* 457–471. https://doi.org/10.1016/j.enpol.2017.03.021

Rabe, B. G. (2022). Racing to the top, the bottom, or the middle of the pack? The evolving state government role in environmental protection. In N. J. Vig, M. E. Kraft, & B. G. Rabe (Eds.), *Environmental policy: New directions for the twenty-first century* (pp 35–62). CQ Press.

Rokeach D., & Schatz, G. (2012). From subsidies to markets: Pursuing a more effective American energy policy. *Renewable Energy Law & Policy, 3(3),*187–195. https://www.jstor.org/stable/24324765

Skjærseth, J. B., Andresen, S., Bang, G., & Heggelund, G. M. (2021). The Paris agreement and key actors' domestic climate policy mixes: Comparative patterns. *International Environmental Agreements, 21,* 59–73. https://doi.org/10.1007/s10784-021-09531-w

U.S. Department of Energy (DOE). (2023a). *Overview of Office of State and Community Energy Programs (SCEP) resources.* https://www.energy.gov/sites/default/files/2023-12/SCEP-BIL-IRA_Fact-Sheet-FY23.pdf

U.S. Department of Energy (DOE). (2023b). *Communities LEAP (Local Energy Action Program) pilot.* https://www.energy.gov/sites/default/files/2023-05/Communities-LEAP-Fact-Sheet-April-2023.pdf

U.S. Energy Information Administration (EIA). (2024). *Renewable energy explained: Portfolio standards.* https://www.eia.gov/energyexplained/renewable-sources/portfolio-standards.php

Vivadili, N., Suleymanov, E., Bulut, C., & Mahmudlu, C. (2017). Transition to renewable energy and sustainable energy development in Azerbaijan. *Renewable and Sustainable Energy Reviews, 80,* 1153–1161. https://doi.org/10.1016/j.rser.2017.05.168

Wesseh, P.K., & Lin. B. (2015). Renewable energy technologies as beacon of cleaner production: A real options valuation analysis for Liberia. Journal of Cleaner Production, *90,* 300–310. https://doi.org/10.1016/j.jclepro.2014.11.062

Wiser, R., Barbose, G., & Holt, E. (2011). Supporting solar power in renewable portfolio standards: Experience from the United States. *Energy Policy, 39,* 3894–3905. https://doi.org/10.1016/j.enpol.2010.11.025

Zhu, J. D. (2021). Cooperative equilibrium of the China-US-EU climate game. *Energy Strategy Reviews, 39,* 100797. https://doi.org/10.1016/j.esr.2021.100797

DISCUSSION QUESTIONS

1. How do we define energy transition? What are the problems of the current energy system? What benefits are expected if the energy transition in favor of renewables and away from fossil fuels is successful?

2. How does RPS differ from FITs? What are the strengths and weaknesses of RPS toward energy transition? As RPS has been a major renewable energy policy tool in the United States with a huge variance by states, what are the challenges of implementing RPSs in the US?

3. How could energy transition affect socioeconomically disadvantaged groups that have been marginalized, underserved, and overburdened by pollution? How could the energy transition be made just and inclusive in consideration of those socioeconomically disadvantaged groups?

UNIT II

Contemporary Issues and Toward a Sustainable World

This unit discusses and addresses contemporary environmental issues as they gear toward a sustainable world. First, in the chapter on "Public Participation and Environmental Justice: Access for Federal Decision Making," Daley and Reames discuss the U.S. federal agencies' approach to including the public, particularly historically disadvantaged and marginalized groups, in agencies' environmental decision making. Environmental justice has two components: distributional justice and procedural justice. This chapter underscores the significance of achieving procedural justice in environmental decision making. The participatory processes in three agencies, the Environmental Protection Agency, the Department of Energy, and the Department of Transportation, were analyzed.

The second chapter, "Mushroom Packages: An Ecovative Approach in Packaging Industry," presents creative destruction for sustainable packaging using an example of Ecovative, a start-up company with technical competency, systems-thinking skills, and stakeholder engagement capacity. This chapter reports the serious concerns about voluminous and hard-to-be-decomposed packaging product-based wastes. After that, it discusses how companies pursuing ecological sustainability can innovate their products using bio-based materials. The Schumpeterian creative destruction contests the idea of steady improvement in addressing wastes and infers how society could pursue product innovation and which policy approaches could support sustainable innovation for companies, as demonstrated in Ecovative.

The third chapter, "The Inclusive City: Urban Planning for Diversity and Social Cohesion," presents the critical role of inclusion and diversity in urban sustainability, expanding the notion of sustainability beyond economic and ecological dimensions. Schreiber and Carius describe the trend of urbanization in developing and emerging economies due to the rural-urban migration. The megacities, such as Cairo, Jakarta, Lagos, and Mumbai, illustrate the twofold process of urbanization; cities are growing in population and becoming increasingly diverse and ethnically heterogeneous. This presents enormous challenges, as cities need to manage the integration of their new migrants into society and urban life, as well as ensure continued social cohesion. The chapter elicits needed, timely interest in social cohesion as climate migrants grow and add to cities in developing and emerging economies with low urban planning capacity. Urban sustainability needs to consider how to make growing cities more inclusive.

The fourth chapter, "Renewable Energy Delivery and Expansion with Public and Private Partnerships for the Global South," presents a unique case study of addressing energy poverty in emerging economies. As 24-hour energy access has not been defaulted in some developing countries, public and private partnerships for renewable-based electricity generation could be a powerful vehicle to grow the economy with stable energy provision while contributing to reductions in greenhouse gas emissions. The case of Gujarat in India underscores the green incentives designed to benefit local residents who permitted the use of their rooftops for electricity generation. The chapter also showcases how federal policy directives can be nicely woven into state and local governments' cooperation for policy implementation.

The final chapter, "Carbon Markets: Principles and Current Practices," examines the theoretical framework of a cap-and-trade program in which companies could flexibly and cost-effectively meet their carbon emissions targets. Carbon markets have arisen from the EU Emissions Trading System and now include regional, national, and state-level cap-and-trade systems and carbon taxes. At the time of writing this Unit II introductory piece, the World Bank reports that there are around 73 carbon schemes in which carbon permits are priced from $0.82 to $155.87. Carbon markets illustrate how market-based mechanisms help reduce carbon emissions, as they can drive companies to compare their costs of reducing carbon inside the firms and carbon prices in marketplaces. By choosing either to reduce inside or buy/sell carbon permits, companies can minimize their carbon abatement costs while complying with mandatory or voluntary carbon targets. Uniformity and connectivity across different markets are the next step toward a sustainable future, but carbon markets symbolize how international climate treaties could trigger county-level policy responses and market reactions to cope with climate change.

In sum, environmental justice, social innovation through corporate sustainability, urban planning, and sustainability in consideration of diversity and inclusion, public-private partnerships (PPPs) for renewables, and carbon markets are all illustrative of emerging environmental issues and critical approaches that can replace traditional prescriptive regulations. In doing so, human society can address environmental, climate, and energy challenges more effectively and overcome collective action dilemmas through which all social actors cooperate to promote environmental public goods and create a sustainable future.

Public Participation and Environmental Justice

Access to Federal Decision Making

By Dorothy M. Daley and Tony G. Reames

Executive Order 12898 on Environmental Justice (EO 12898) directs federal agencies to develop broad strategies to identify and address any disproportionate, negative impacts stemming from agency activities and affecting minority and low-income communities. It aims to integrate environmental justice considerations into the standard operating procedures of federal agencies, and in doing so, mitigate the likelihood that minority and low-income communities experience concentrated environmental burdens. By their very nature, environmental justice concerns are not uniformly distributed across the country. Rather, they tend to be local or regional in scale, and the specific details of the environmental justice situation may be unique. For example, contaminated drinking water in the Appalachians, poor air quality in Southern California, or traffic congestion in Houston may disproportionately affect low-income communities, minority communities, or both, but these problems occur on a local or regional scale. Therefore, a "one size fits all" approach to environmental justice is unlikely to be efficient or effective. The diversity and distribution of environmental justice problems pose a considerable challenge to federal agencies charged with implementing the executive order.

Public participation is one mechanism to address this challenge. Effective public participation could not only help identify and characterize environmental justice concerns, but it could also inject critical local knowledge and inform policies and programs to solve environmental justice problems (Dietz et al. 2008). EO 12898 emphasizes the need for widespread public participation to better identify, understand, and tackle environmental justice concerns. Consequently, the definition of environmental justice by the Environmental Protection Agency (EPA) underscores the importance of public participation by calling for "meaningful involvement of all people" in every aspect of environmental decision making.

At the federal level, cultivating and maintaining widespread and diverse public participation to prevent and combat concentrated environmental risk requires significant agency commitment (Dietz et al. 2008; Innes and Booher 2004; Foreman 1998). Although EO 12898 tasks federal agencies to increase public participation, to date, there has been limited evaluation aimed at understanding how—after twenty years of implementation—agencies have addressed that charge. This chapter evaluates the ways in which federal agencies utilize public participation to address environmental justice in minority and low-income communities in response to EO 12898.

The chapter begins by examining the research on public participation and environmental decision making in general. Based on this, we highlight the opportunities and challenges of using participation to address environmental justice problems. To better understand how EO 12898 has influenced public participation in federal agencies, we examine the EPA, the agency responsible for leading the federal government's approach on environmental justice. In addition, the chapter compares and contrasts how the EPA, the Department of Energy (DOE), and the Department of Transportation (DOT) utilize public participation to advance environmental justice. Our assessment of these agencies is based upon examining a range of government documents and electronic sources that detail each agency's environmental justice strategy in 1995 and subsequent revisions. In addition, we compiled annual reports from the three agencies, along with any existing evaluations of agency efforts related to implementing the executive order. Material from these sources is integrated with findings from existing research on civic engagement, public participation, and social equity to provide a general assessment of how well each agency has expanded opportunities for public participation in light of the executive order. Finally, the chapter concludes with policy prescriptions for federal agencies as they move forward with their reaffirmed commitment to environmental justice and public participation.

Public Participation and Environmental Decision Making

Public participation is woven into the fabric of American democracy and it can take a variety of forms. Voting, letter writing, attending public meetings, forming an interest group, or serving on advisory committees are all examples of public participation. In recent decades, public participation has been an increasingly important element of environmental governance (Bulkeley and Mol 2003; Daley 2013; Dietz et al. 2008). Public participation in environmental decision making can be *process-oriented*. For example, public comment periods for permitting, setting standards, and writing and revising regulations are all avenues to engage the public in the process of environmental decision making. In addition, when the EPA or other agencies are in the midst of a particularly controversial environmental issue, they may hold listening sessions, workshops, and seminars to provide additional opportunities to gauge public sentiment on potential changes in advance of a final decision. Public participation can also be more *outcome-oriented*. This type of participation tends to focus on existing environmental hazards resulting from past decisions. Examples include protesting the existence of multiple polluting facilities within one community or serving on a community advisory board for a hazardous waste site that needs to be remediated. Outcome-oriented participation can also advocate for strict enforcement and compliance with existing environmental laws by reporting facilities that might be in violation of those laws.

Environmental decisions tend to be highly technical, complex, and value-laden. Scientific and technical expertise is critical to foster better understanding of the nature of environmental problems and potential solutions. But even when scientific and technical knowledge is combined to inform environmental decisions, conflict and gridlock can remain, particularly when public preferences on the nature and distribution of environmental risks and benefits are poorly understood or not carefully considered (Dietz et al. 2008; Weber 2003; Webler and Tuler 2006). Public participation provides an avenue to gauge public preferences and insert public values into environmental decision making. Recent research suggests that when done well—an important caveat—public participation can increase equity, reduce conflict and gridlock, and lead to improved environmental decision making (Dietz et al. 2008; Klyza and Sousa 2013; Pellow and Brulle 2005). There are a number of ways that public participation could result in these desirable outcomes. Widespread and meaningful participation can add legitimacy to any final decision and enhance overall levels of trust in government. Local knowledge could provide more detailed information on the nature of a problem, along with identifying locally appropriate solutions that capitalize on community norms and in doing so, increase the chances of successful policy implementation (Bulkeley and Mol 2003; Daley 2013; Dietz et al. 2008; Reed 2008).

Conversely, when done poorly, public participation can reduce overall levels of trust in government, increase conflict and gridlock, and highlight weaknesses in both the decision-making process and outcomes from that process. Thus, the promise of public participation hinges on both the quality of the participatory process and the overall ability of stakeholders to engage in the process. Effective or high-quality participatory processes tend to have a clear goal, adequate financial and human capital, consistent institutional commitment, an ability for participatory action to influence all stages of decision making, and an emphasis on implementation and evaluation (Beierle and Cayford 2002; Charnley and Engelbert 2005; Dietz et al. 2008; Leach 2006; Reed 2008; and Sirianni 2009).

Most environmental legislation calls for some form of public participation, and EO 12898 is no exception. Traditional participatory mechanisms used in environmental legislation include public hearings and public comment periods for permits, regulations, and other environmental decisions. One-time seminars and workshops, occasional roundtables, and advisory committees are increasingly being utilized to improve the scope and impact of public participation (Innes and Booher 2004; Kellogg and Mathur 2003; Spyke 1999; Beierle 1998; Coglianese 1997). Although some of these approaches to public participation can be more innovative than others, research suggests that as currently implemented, all of these mechanisms tend to result in top-down, expert-driven decision making, despite consistent calls for more bottom-up approaches to public participation (Agyeman and Angus 2003; Dietz et al. 2008; Laurian 2007). Therefore, in their present form, these participatory mechanisms would not be *effective* participatory processes to reduce conflict and gridlock or improve equity (Beierle 1998; Brion 1988; Buchanan 2010; Fine and Owen 2005; Hernandez 1995; Webler and Tuler 2006).

In many instances, both the quality of the participatory process and the ability of stakeholders to engage in that process are compromised (Klyza and Sousa 2013). Determining who represents the affected public and how to broaden the scope of public participants in a decision-making process remains difficult. Government agencies commonly recruit people from organized groups who have expressed interest or are active around a particular issue (Larson and Lach 2008). Although this seems like an efficient strategy, it raises concern about the degree to which organized interests represent the views of a broader community. Relying on this approach to insert public values into decision making may not expand access to new participants, as it tends to result in a similar set of engaged stakeholders taking part in the decision-making process (Coglianise 2006; Agyeman and Angus 2003; Beierle 1998). Comparatively, business interests are much easier to identify and engage in the decision-making process. They have considerable incentive to remain involved in environmental decision making, particularly when decisions directly regulate their behavior. Moreover, these concentrated interests tend to have more resources than other stakeholders, and they are in a better position to shape environmental decision making in their favor (Furlong 2007; Kamieniecki 2006; Yackee and Yackee 2006).

Federal agencies face a complicated decision-making environment: While regulated interests are motivated and likely to remain active in any and all aspects of decision making, engaging the public is far more challenging. Overall levels of civic engagement and public participation in the United States are declining (Macedo et al. 2005). Inconsistent public participation directly limits the ability of this tool to give voice to a diverse public (Dietz et al. 2008; Foreman 1998). Meaningful public participation within federal agencies requires creativity, resources, and a consistent commitment to a participatory process. Since the Reagan era, the federal government has experienced near-constant pressure to devolve responsibilities to state and local government whenever possible. While devolution to subnational governments may provide more opportunity for public participation in state and local institutions (Macedo et al. 2005), it significantly complicates the ability of federal agencies to maintain well-resourced public participation processes.

Public Participation and Environmental Justice: Opportunities and Challenges

Successfully incorporating public participation into any environmental decision-making process remains a significant opportunity and a challenge. These opportunities and challenges are magnified with environmental justice concerns. Because public participation can increase equity and reduce conflict and gridlock in decision making, it holds tremendous promise for both ameliorating existing environmental justice problems and preventing future concentrated risks. However, this promise is situated in a broader context of racial, economic, and political inequality (Brulle and Pellow 2006; Cole and Foster 2001; Schlozman, Verba, and Brady 2012). Minority and low-income communities are less politically active overall (Rosenstone and Hansen 1993; Schlozman, Verba, and Brady 2012), and historically, they have mobilized less around environmental issues than wealthy, white communities have (Brulle and Pellow 2006; Gauna 1995). The modern environmental movement is largely comprised of white, middle class Americans concerned with natural resource degradation (Baber and Bartlett 2013; Gauna 1995); often, social justice and equity are not on the agenda (Mohai, Pellow, and Timmons Roberts 2009). In contrast, the environmental justice movement is distinctly shaped by the history of civil rights in the United States, as well as grassroots anti-toxics activism stemming from environmental disasters like Love Canal. As a result, there tends to be more outcome-oriented public participation in minority and low-income communities that focus on existing environmental risks in their communities.

Some within the environmental justice community view the disproportionate impact of environmental hazards on minority and low-income communities as a product of the same social structure that produces racial oppression. Others view the concentration of burdens as an intended consequence of dominant economic forces (Brulle and Pellow 2006; Cole and Foster 2001; Foreman 1998). Regardless, either of these causal stories creates significant hurdles in using public participation to advance equity in environmental decision making. Successful public participation in environmental decision making is predicated on trust. As with any collaborative relationship, high levels of trust facilitate productive interactions (Lubell et al. 2005). When that trust is frayed, the promise that public participation would increase equity and reduce conflict and gridlock in decision making becomes more difficult to realize.

Despite this difficulty, public participation can provide critical information regarding the ways in which local residents experience concentrated environmental burdens. From a pragmatic point of view, the expansive geography of the United States and the local or regional scale of environmental justice concerns combine to highlight the need for public participation as an important mechanism to advance environmental justice in federal agencies. But patterns of public participation directly affect the ability of this tool to ameliorate environmental justice concerns. In fact, some contend that public participation in a decentralized political system actually contributes to environmental justice problems (Foreman 1998; Gauna 1995; Munton 1996).

Local grassroots mobilization, described as "Not In My Backyard" or NIMBY, has been a powerful response to environmental risks (Fletcher 2003; Kraft and Clary 1991). Over the years, politically active communities have mobilized successfully to block power plants, hazardous waste sites, landfills, and other locally unwanted land uses (LULUs) from being established in their vicinity. This participatory activity increases the likelihood that LULUs are instead located in communities with lower levels of political mobilization and public participation, which often are minority and low-income neighborhoods. NIMBY can extend beyond siting; local groups may advocate stringent enforcement of laws governing existing facilities within their communities and demand swift action in response to accidents (Gauna 1995), while less organized communities receive less attention. Thus, this type of grassroots activism can facilitate the concentration of environmental risks in minority and low-income communities in multiple ways.

The past several decades have witnessed a consistent increase in opportunities for public participation in almost all aspects of environmental decision making. However, there is little evidence of a corresponding increase in widespread, diverse public involvement. In fact, evidence suggests that levels of civic engagement and public involvement in the United States have declined (Coglianese 2006; Putnam 2000; Sander and Putnam 2010), and existing patterns of participation tend to represent white, wealthy, and educated citizens (Macedo et al. 2005; Schlozman, Verba, and Brady 2012). Minority and low-income communities may face higher barriers to entry for public participation. Public participation requires some general understanding of how agency decisions affect citizens' interests. Lower levels of educational attainment, language barriers (Larson and Lach 2008; Fine and Owen 2005), and limited knowledge about federal decision-making processes or how decisions affect their lives are likely to constrain citizens' motivation to participate (Coglianese 2006).

Despite these challenges, public participation remains an important tool in advancing environmental justice considerations within federal agencies. In 2011, the seventeen agencies covered under EO 12898 declared, "the continued importance of identifying and addressing environmental justice considerations in agency programs, policies and activities"[1] by signing a Memorandum of Understanding on Environmental Justice and Executive Order 12898.

The EPA, Public Participation, and Environmental Justice

The EPA is the one of the most active federal agencies in terms of incorporating environmental justice into agency activities, and it is also responsible for convening and leading the Interagency Working Group on Environmental Justice. Despite a concerted effort to integrate environmental justice and public participation into agency activities, progress remains uneven (GAO 2011; CCR 2003). In part, this reflects the nature of the participation challenges outlined above, as well as highlighting the challenge of changing bureaucratic behavior. The EPA is a large bureaucracy charged with implementing a wide variety of environmental legislation, including the Clean Water Act (CWA), the Clean Air Act (CAA), the Resource Conservation and Recovery Act (RCRA), the Toxic Substance Control Act (TSCA), and several others, in addition to EO 12898. The agency's duties are scientifically complex and administratively difficult, with decision making often generating conflict across a range of interests, including environmental justice stakeholders. Environmental justice issues may arise in a number of activities for which the agency is responsible, including setting standards, permitting facilities, awarding grants, issuing licenses and regulations, and reviewing proposed actions by other federal agencies.

Public participation is a key element underpinning the EPA's environmental justice strategy as the agency seeks to "work with communities through communication, partnership, research and the public participation process" and also "help affected communities have access to information which will enable them to meaningfully participate in activities" (EPA 1995, p. 3). Shortly after EO 12898 was signed, the agency acknowledged that a host of stakeholders are directly affected by the agency's environmental decisions and therefore "must have every opportunity for public participation in the making of those decisions" (EPA 1995, p. 4). Environmental justice problems tend to be local in nature; therefore, participation and outreach efforts need to be tailored to the specific problem and community context. But this type of customization is complicated. In the first ten years following EO 12898, EPA regional offices relied on different approaches to identifying minority and low-income communities; this variation created significant implementation and evaluation hurdles (EPA 2004). As research on public participation suggests and the EPA's experience highlights, identifying the affected community is challenging (Daley 2013; Dietz et al. 2008; Larson and Lach 2008), and absent this identification, public participation is not likely to yield increases in equity.

Environmental justice staff within the agency have a challenging mission: to integrate environmental justice considerations into all aspects of a major bureaucratic organization and create systemic change, including increasing opportunities for meaningful public participation. Historically, the EPA has been a "stove-piped" organization with separate offices for air, water, and toxics, each operating largely in their own silos. The agency has no overarching legislative agenda, but rather a series of diverse environmental problems to address. Without clear guidance on how to prioritize among divergent environmental efforts, the agency has been criticized as engaging in turf battles (Rosenbaum 2002). Creating meaningful change in a large hierarchical organization requires significant time (up to a decade or longer), consistent commitment from all levels of personnel, considerable resources, and multiple, overlapping strategies (Bardach 1998; Brehm and Gates 2002; Mazmanian and Sabatier 1989; Wilson 1989).

In November 1992, an Office of Environmental Equity (later renamed the Office of Environmental Justice (OEJ)) was created within the agency, and each regional office has environmental justice coordinators to integrate environmental justice into its policies, programs, and activities. Over the last two decades, the agency has relied upon a range of participatory mechanisms to engage minority and low-income communities in environmental decision making. These include public meetings, public notice and comment periods, developing partnerships, conducting seminars and workshops, and convening the National Environmental Justice Advisory Council (NEJAC). Since 1993, NEJAC has been providing independent advice and recommendations to the agency on environmental justice issues. The approximately 26 members of the committee represent stakeholders from academia, business and industry, state and local governments, tribal governments, environmental organizations, community groups, and nongovernmental organizations.

Agency and NEJAC documents note the importance of public participation, and several also highlight successful collaborative relationships, but it is difficult to determine if these success stories are representative of the norm (EPA 1997). Similarly, it is difficult to determine if EPA's general approach to participation has consistently resulted in increased diversity of participation. In 2006, the agency released a series of environmental justice accomplishment reports on their website to highlight environmental justice integration, including public participation, into the agency's decision-making processes.[2] Notably, two (out of ten) regional offices and three (out of nine) offices within the agency's headquarters do not provide accomplishment reports. The available reports provide detailed tables listing the region's or office's activity, output, outcome, and results in relationship to a set of goals and objectives. They provide an excruciating level of detail, but this actually makes understanding the overall progress toward integration more difficult.

For example, one objective from the Office of Air and Radiation (OAR) report included providing opportunities for meaningful involvement between agency staff and affected community members. OAR listed six activities they developed to achieve this goal, one of which was the development of training modules to facilitate public participation in air quality permitting in environmental justice communities. Another activity included year-round monitoring and reporting of the Air Quality Index (AQI). The training module to enhance engagement in air quality permitting was not funded and therefore had no impact on participation. But year-round monitoring and reporting of the AQI led to the development of an e-alert system. This allows individuals to receive electronic air quality information, which can be particularly useful for limiting exposure to air pollution. The report does not indicate if the agency engaged environmental justice communities to encourage participation in the e-alert program. They may have done so, and it is simply not listed in this report. Absent more context about which activities are critical in achieving meaningful involvement, it is hard to evaluate the accomplishments listed in these reports.

Along with using traditional participatory approaches such as public notice and comment periods, the agency has invested considerable effort in capacity building by improving access to environmental and demographic information. Providing information to affected communities and interested citizens is one mechanism to level the playing field and facilitate effective participation. EJScreen, for example, is an interactive mapping tool on the EPA's website (https://www.epa.gov/ejscreen) which allows users to see how environmental burdens may be concentrated within a geographic area. Any facility reporting to EPA can be identified on a map, along with water-monitoring stations. This includes current hazardous waste transfer, storage and disposal facilities, Superfund sites, Brownfield areas, Toxics Release Inventory (TRI) facilities, and air and water dischargers. This tool overlays demographic information as well, making it relatively easy to identify if low-income or minority communities have a large number of EPA-permitted facilities compared to wealthy, white communities. EJScreen also provides critical health outcomes, such as information on infant mortality rates, birth rates, and cancer risk. This is a powerful diagnostic tool that communities and advocacy groups could use to identify potential environmental justice issues. In its current form, however, it falls short. While the agency has devoted considerable resources to create this level of access to information, EJScreen lacks any corresponding information about how to *act* on this information. This is a missed opportunity for the agency to facilitate participation. If people use EJScreen to better understand the density of permitted facilities within their community, what next steps could they take to engage in environmental decision making? Currently, there is no guidance connecting the information provided in EJScreen to information on the agency's permitting process, rulemaking, or other elements of environmental decision making.

In comparison, the EPA's Community Action for a Renewed Environment (CARE) program exemplifies a model initiative to provide information and empower communities while also facilitating public engagement (Hansell, Hollander, and John 2009; Sirianni 2009; EPA 2011). This competitive grant program aims to reduce toxic pollution by (1) working collaboratively with local communities to reduce exposure to toxic pollutants, (2) helping communities understand individual and cumulative sources of toxic exposure, (3) working with communities to identify and prioritize risk-reducing activities, and (4) creating long-term, community-based partnerships to improve and protect the local environment.[3] While many federal agencies, including the EPA, struggle with how best to work with local communities, CARE demonstrates that national-local partnerships can be effective and have considerable spillover benefits. For example, two different CARE communities identified exposure to solvents and other chemicals used in auto body shops as an environmental risk that could—and should—be minimized. On the local level, environmental groups, the EPA, and local auto body shops worked collaboratively to identify alternative, less toxic chemicals for paint stripping, along with exploring opportunities for improved disposal practices. This experience, in turn, prompted the EPA to work closely with these CARE grantees to develop national emission standards regulating paint strippers and other solvents used in auto body refinishing (Hansell, Hollander, and John 2009). These regulations broaden the impact of the CARE program because they apply to auto body shops across the country.

The focus of the CARE program often results in investment in environmental justice communities, but any community can apply for support from this program. The Environmental Justice Small Grants program, administered by the EPA since 1994, is specifically designed to support environmental justice communities. These grants provide a tangible way to identify local issues, support community groups, and build trust between the agency and affected community members. Reflecting the diversity of environmental justice concerns, grants have focused on a wide range of environmental issues, such as radon, lead, farmworker safety, recycling, water quality, and children's health.[4] But the commitment to investing in communities has varied greatly over time. Figure 2.1.1 shows the number of Environmental Justice small grants awarded by year. The number of grants peaked shortly after EO 12898 was signed, and while there were increases around the time of the American Recovery and Reinvestment Act, more recent years have witnessed a decline. It is not possible to conclude if the variability in grants awarded reflects drastic changes in the applicant pool, changes in agency priorities, or is simply the result of federal budgetary constraints.

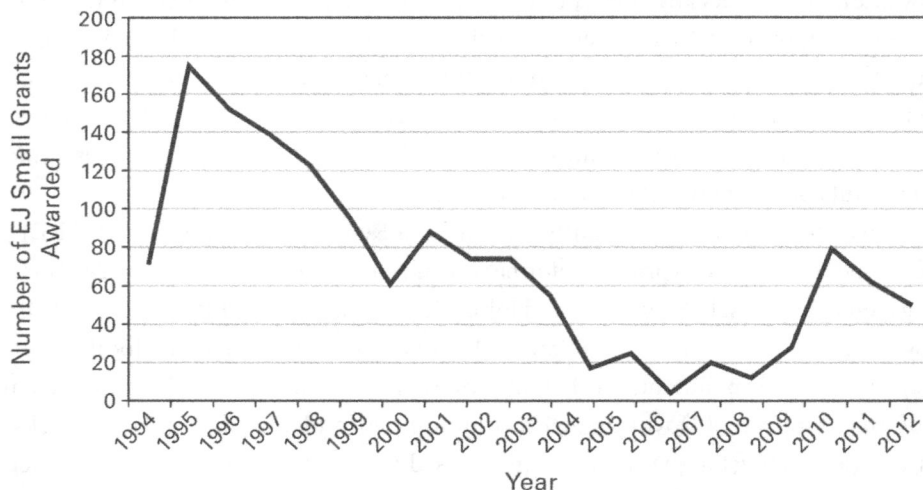

Figure 2.1.1 Number of Environmental Justice (EJ) Small Grants Awarded by Year (1994–2012)
Source: Data in this chart are drawn from reports available at http://www.epa.gov/ environmentaljustice/grants/ej-smgrants.html

Moreover, recent research comparing all Environmental Justice Small Grants awarded between 1994 and 2004 suggest some challenges in targeting environmental justice communities. During this time period, the majority of the funds distributed were not directed toward minority and low-income communities with higher than the nation's average TRI emissions (Vajjhala 2010). It is impossible to determine if this pattern is due to a selection effect, meaning that those minority and low-income communities with above-average TRI emissions did not apply for a grant, or if in fact, there are problematic patterns in the distribution of funds. The Environmental Justice Small Grants program could be an innovative tool that builds trust and increases participation, but inconsistent support or biased implementation will limit its reach; effective participatory mechanisms require consistent commitment.

Environmental justice problems are diverse and may require that the EPA devote a range of expertise within its agency to understand and address community concerns. Agency leadership and commitment to high-quality participatory practices are necessary to tackle most environmental justice problems. Many in the environmental justice community worried that in 2001, under President George W. Bush, that EO 12898 would be repealed. This, in fact, did not occur. But the change in presidential administration did result in a shift in emphasis within environmental justice at the EPA. Christine Todd Whitman, the first administrator of the EPA under President Bush, issued a memo restating the agency's commitment to environmental justice. Rather than highlighting the significance of minority and low-income communities in environmental justice, the agency opted to focus on environmental justice for everyone (EPA 2004; Mohai, Pellow, and Timmons Roberts 2009). This change in focus, deemphasizing minority and low-income communities, adds additional challenges to expanding meaningful access for public participation within agency decision making. It strains trust in public institutions with stakeholders who were likely already suspicious of the federal government's commitment to social equity and environmental justice.

Under President Barack Obama, the EPA has renewed its focus on environmental justice and reiterated the importance of engaging minority and low-income communities to achieve environmental justice. In 2010, EPA administrator Lisa Jackson listed environmental justice as one of the agency's top priorities, and the following year, the agency published its blueprint to achieve environmental justice, *Plan EJ 2014*. This plan is an innovative strategy, and its structure suggests that agency leadership have clearly received the message: agency fragmentation is a significant hurdle in advancing environmental justice. *Plan EJ 2014* emphasizes cross-agency responsibility for advancing environmental justice into all aspects of environmental decision making. The plan also aims to improve the scientific and technical tools available to better diagnose and prevent the concentration of environmental risk in any community, along with evaluating more carefully the way that the agency's current initiatives can be tailored to support environmental justice goals.

Public participation looms large in this plan; the three main goals include to "protect the environment and health in overburdened communities; empower communities to take action to improve their health and environment; and establish partnerships with local, state, tribal, and federal governments and organizations to achieve healthy and sustainable communities." (EPA 2011, p. 1). *Plan EJ 2014* highlights the need to strengthen community-based programs with a particular focus on identifying "scalable and replicable" programs to more effectively address concentrated environmental burdens. For example, successful programs like CARE should be replicated and expanded whenever possible, and as the agency proceeds with implementing *Plan EJ 2014*, it intends to evaluate its progress regularly, along with producing information on "lessons learned" to facilitate communication about strategies that are effective in addressing environmental justice problems.

Some of the more innovative participatory approaches have emerged in recent years as the agency reinvigorates its commitment to environmental justice. In advance of developing *Plan EJ 2014*, the agency held a series of community forums and listening sessions across the country. It has also integrated communication technology to expand public participation, holding quarterly community outreach teleconference calls to gather and respond to community concerns. The OEJ hosts these calls, and agency staff from other programs and offices are present to respond to concerns that were submitted beforehand, as well as answer questions from the public during the call. Once a call is completed, the

EPA provides both an audio version and a written transcript of it on the agency's website. Putting aside any consideration of the digital divide, this type of community outreach has the ability to expand access to a wide range of interested citizens. Recently available transcripts suggest that fragmentation within the agency and between the EPA and its state counterparts remains frustrating for community members trying to engage in environmental decision making. One caller noted that existing rules from the Superfund program, combined with longstanding agreements between the EPA and a state environmental agency, were contributing to the concentration of drinking water risk in a community, rather than protecting community members and providing a platform for a safe environment.[5]

While a renewed commitment to environmental justice is clear in agency documents, advancing a streamlined collaborative approach in a large organization remains difficult. *Plan EJ 2014* holds tremendous promise for advancing environmental justice. With an emphasis on cross-cutting implementation and improved access to information, *Plan EJ 2014* could enhance the diversity of public participation in environmental decision making. It aims to strike a balance between creating some standard operating procedures—like a national approach to identifying environmental justice communities—and providing discretion and flexibility to ensure that solutions are tailored to local problems.

As with all innovative plans promising change, the devil is the implementation details (Mazmanian and Sabatier 1989; Pressman and Wildavsky 1984). In the coming years, the challenge within EPA includes maintaining this renewed focus on environmental justice. Working across the breadth of programs within the agency and with stakeholders outside the agency to more successfully integrate public participation and environmental justice into the agency's decision-making processes will require resources and dedication. And, although the agency has had some success in expanding the conversation, improving access to information, and forging new partnerships with communities, more work remains, as the scope of environmental justice problems is large and public participation provides one potential mechanism for successful resolution of conflict and improved decision making.

Federal Agencies, Public Participation, and Environmental Justice

While the EPA is a lead agency in advancing environmental justice within the federal government, it is by no means the only federal agency that struggles to integrate EO 12898 (including its focus on public participation) into daily routines. To compare and contrast the approach of other federal agencies to the EPA, we examine environmental justice and public participation in two other agencies: the DOE and the DOT. Table 2.1.1 compares the original participatory goals for the three agencies immediately after EO 12898 was signed, along with current participation goals. Material for this table is drawn exclusively from the government documents cited in the text. The table also highlights some of the participatory approaches commonly used by the agencies to address environmental concerns.

Table 2.1.1 Comparison of Environmental Justice and Public Participation across Federal Agencies

Agency	EPA	DOE	DOT
Agency mission	Protect human health and the environment.	Ensure the nation's security and prosperity using science and technology to address energy, environmental, and nuclear challenges.	Ensure fast, safe, efficient, accessible, and convenient transportation systems.
Public participation goal (1995 EJ strategy)	To achieve widespread opportunity for participation in environmental decision making. An informed and involved community must underpin environmental protection.	Improve DOE's credibility and trust by enhancing public participation in agency activities.	Bring government decisions closer to the communities affected by them; expand opportunities for public participation in decisions relating to human health and the environment.
Public participation goal (updated strategy)	Public participation underpins two of the three main goals in *Plan EJ 2014*.	In more recent documents, the DOE's public participation goals remain unchanged.	To ensure the full and fair participation by all potentially affected communities in the transportation decision-making process. (DOT 2012 EJ strategy)
Participatory tools used	• Notice and comment periods • Advisory boards (NEJAC and community advisory boards) • EJ small grants • Translated documents and interpreter • Hot lines/reporting links • Conference calls	• Notice and comment periods • Site-specific advisory boards • Partnerships with minority organizations • Cooperative agreements to provide funding to communities	• Notice and comment periods • Seminars/ workshops • Directs states, MPOs, and transit operators to engage citizens in project planning

The Department of Energy

The DOE's mission is to "ensure America's security and prosperity by addressing its energy, environmental, and nuclear challenges through transformative science and technology solutions."[6] DOE has approximately 16,000 employees based around the country at its various offices, laboratories, and field sites. Among other duties, DOE is responsible for cleaning up the environmental legacy from the nation's nuclear weapons program. This is an extraordinarily technically complex program that must confront very long time horizons, as the half-life on some of the spent nuclear material will remain radioactive for many decades to come. DOE is responsible for more than 80 federal facilities throughout the country that are contaminated and are currently undergoing environmental cleanup. In addition to developing its own environmental justice strategy, EO 12898 tasks the Secretary of Energy or a designated officer to serve on the Interagency Working Group.

Interestingly, DOE's public participation goal is to "enhance the credibility and public trust of the department by making public participation a fundamental component of all program operations, planning activities, and decision making" (DOE 1995). While the strategy emphasizes community participation and empowerment, it seems to suggest that this is done to improve the agency's profile, as opposed to improving or resolving conflict between the department and its stakeholders. Indeed, when the department first published its environmental justice strategy in 1995 following EO 12898, it noted that no stakeholder comments were included in the document because the department did not provide adequate time for public review prior to publishing the strategy.[7] Certainly, this is not an ideal way to establish positive relationships with stakeholders (Dietz et al. 2008). To its credit, DOE noted in early documents that the department's environmental justice strategy is a living document, and there would be ample opportunity for stakeholder contribution and change in the future.

Improving trust and credibility through participation may be the part of the department's approach to environmental justice, but it has not been easy. DOE acknowledges that minority, low-income, and tribal communities have traditionally lacked access to information and technical advisers to be informed participants in complex environmental decisions. NEJAC released a report noting, among other things, the strained relationships between DOE and the affected communities near the federal facilities the department manages (NEJAC 2004). National security and confidentiality issues surrounding activities at DOE facilities directly hamper the ability of the department to build trust and engage stakeholders. Despite this challenge, NEJAC recommended continued investment in public participation.

Currently, the Office of Legacy Management within DOE provides leadership for environmental justice programming. Environmental justice concerns may arise in relation to the treatment, storage, and disposal of hazardous, radioactive, and mixed waste. DOE is responsible for a number of federal sites throughout the nation that are currently undergoing cleanup processes, such as the Hanford site in Washington and the Savannah River site in South Carolina. Environmental justice issues may also arise during the National Environmental Policy Act (NEPA) planning process when the agency is upgrading existing energy projects or constructing new facilities.

DOE utilizes typical participatory mechanisms to engage the public, including public notice and comment periods, along with relying on citizen advisory groups and site-specific advisory boards. The department attempts to ensure that advisory boards reflect the communities they represent, but identifying and engaging an affected community can be difficult. For example, the Nevada Test Site Advisory Board has lacked broad representation for decades. The department attempted to recruit a wide cross section of potential stakeholders, but has had little success. In 2010, the Nevada Site office conducted a membership recruitment drive, advertising in the Las Vegas valley, rural communities, and a Spanish-language newspaper; however, no applications for the advisory board were received (DOE 2011). On the other hand, the Hanford Advisory Board reflects the diverse viewpoints in the affected community and region, including minorities and members of tribal nations (DOE 2011).

While the EPA relies on small grants to invest broadly in communities and environmental justice solutions, DOE takes a different programmatic approach. Many of their public participation initiatives seem to hinge on building technical capacity in affected communities (DOE 2011). This likely reflects both the department's mission to rely on "science and technology" to meet energy challenges, along with the highly technical nature of the cleanup at DOE facilities. The department supports efforts by historically black colleges and universities to build the science, technology, and engineering workforce and fills a variety of intern positions with minority students. In 2009, DOE entered into fifteen cooperative agreements with tribal nations, totaling approximately $6 million. This financial support is designed to build local communities' capacity to increase participation in DOE decision making (DOE 2011). Communities can use the funds to hire scientific and technical staff to help examine site cleanup strategies in order to educate community members on the more technical nature of DOE's responsibilities, as well as building trust by increasing access and transparency to site documents and the decision-making process.

In some cases, tribal nations have succeeded in affecting the department's decision making through cooperative agreements. For example, at the Hanford site, the inclusion of tribal input resulted in the protection of cultural, religious, and natural resources relating to Gable Mountain (DOE 2009). DOE's Los Alamos Pueblos Project in New Mexico relies on tribal governments to manage pollution monitoring programs and tribal governments also actively participate in generating the Los Alamos and National Laboratory Site-Wide Environmental Impact Statement by reviewing and commenting on the document (EPA 2013; DOE 2008). Despite success in many areas relating to environmental justice, DOE progress reports note that much work remains (DOE 2009; 2011).

The Department of Transportation

The DOT's mission is to "serve the United States by ensuring a fast, safe, efficient, accessible and convenient transportation system that meets our vital national interests and enhances the quality life of the American people, today and into the future."[8] DOT is the largest of the three agencies compared here, with over 57,000 employees across the country. Within DOT, there are thirteen agencies, ranging from the Federal Highway Administration to the Federal Transit Administration to the Pipeline and Hazardous Material Safety Administration, working to accomplish DOT's broad mission. In addition to developing its own environmental justice strategy, EO 12898 tasks the Secretary of Transportation or a designated officer to serve on the Interagency Working Group.

In its 1995 environmental justice strategy, DOT expressed its commitment to "bringing government decision making closer to the communities and people affected by [its] decisions and ensuring opportunities for greater public participation in decisions relating to human health and the environment" (DOT 1995). DOT published its draft strategy in the Federal Register and mailed approximately 3,000 copies to constituent groups and representatives of the environmental justice community. They received approximately 50 comments and modified the final version to incorporate suggestions. In 1997, DOT issued an internal order on environmental justice as key component of the department's strategy.[9] In the internal order, the department reaffirmed its commitment to establishing or expanding procedures that enhance the participation of minority and low-income communities during all stages of the department's decision making. DOT updated its strategy in 2012, stating that its guiding principle on public participation is "to ensure the full and fair participation by all potentially affected communities in the transportation decision-making process" (DOT 2012).

Most of DOT's work involves setting policies and procedures to be followed by state and local governments in planning and constructing transportation facilities that receive federal funding. Stakeholder involvement, as required by DOT guidelines, is often carried out by state and local agencies responsible for the process. Environmental justice issues arise most frequently when some communities get transportation benefits, while others experience fewer, or when some communities suffer disproportionate negative impacts from transportation programs. DOT, like the EPA and DOE, struggles with representation in public participation, as some communities are consistently less well represented than others during policy and decision making concerning transportation resources (Cairns et al. 2003).

In the late 1990s, DOT acknowledged increasing public concerns regarding compliance with environmental justice provisions during metropolitan and statewide transportation planning (DOT 1999). In response, the department requested that administrators within DOT raise a litany of questions during state, metropolitan planning organization (MPO), and transit operator certifications pertaining to equity and public involvement in particular. Recognizing room for improvement in engaging low-income and minority communities in the public participation process, the department directed administrators to review public participation plans for agencies receiving federal funds and to address the strengths and deficiencies (DOT 1999).

In recent years, DOT has released "Environmental Justice Implementation Reports" available on the agency's website.[10] Like EPA and DOE, DOT engages in the typical range of public participation processes, including public notice and comment periods, convening advisory groups, and holding seminars and workshops. DOT is also using technology to increase public participation. In advance of finalizing its most recent environmental justice strategy, the department held traditional public notice and comment periods but added to this approach with a novel electronic participatory mechanism called EJ Ideascale. This electronic comment feature allowed the public to contribute their ideas *and* also view and respond to public comments made by others.[11]

According to the department's website, a considerable number of comments have been received on their updated environmental justice strategy. The department's response to public comments are also available online.[12] Yet again, the public comments highlight the challenges of advancing environmental justice in a large-scale, multifunctional bureaucracy. Some public comments strongly support the need for increased harmonization and integration of a unified environmental justice strategy within the department, while others advocate for more flexibility to ensure that decision making can accurately take into account unique, place-based features. Under these conditions, implementing agencywide environmental justice strategies remain challenging: federal agencies receive virtually no guidance on how to proceed when a diverse public provides conflicting input.

Conclusion

There is tremendous variation across federal agencies; for instance, they have divergent missions, different professional norms, and diverse agency cultures. EPA operates not only through its national offices, but also through its ten regional offices, DOE functions through various offices, laboratories, and field sites, while DOT works through hundreds of state transportation departments, MPOs, and local transit authorities, each of which enjoys substantial decision-making discretion. Such an overlay of arrangements leads to considerable variations in policy approaches across jurisdictions, especially since cities, states, and regions vary widely in their environmental problems. Thus, in many instances, environmental justice must contend with a fragmented and decentralized policy process.

Our assessment is based on examining a range of existing government documents and electronic sources. We compare information from these sources to existing research on civic engagement, public participation, and social equity to examine how well federal agencies meet the participation challenges embedded in EO 12898. Overall, the environmental justice movement and this executive order have increased opportunities for public participation in federal decision making. But despite the increased opportunity, participation from minority and low-income communities remains uneven. Perhaps this should not be surprising, given what we know about both the challenges inherent in generating diverse participation, particularly in communities that have been historically disenfranchised, and the challenges of creating systemic change in large bureaucracies.

The three federal agencies examined here have all experienced some success in using public participation to address environmental justice concerns, and they have also faced significant challenges.

These challenges stem from multiple sources: (1) a complex decision-making environment at the federal level, complicated by devolution and privatization; (2) the capacity of large public organizations to incorporate public participation is variable; (3) the technical nature of environmental problems, which demands some level of competence on the part of the affected public, and the capacity of citizens to engage in scientific and technical decision making; and (4) the general decline in public engagement over time.

While there is significant promise in using public participation to address environmental justice concerns, this promise hinges on developing and maintaining effective public participation mechanisms. Prior research highlights that effective participation is characterized by clear goals, adequate human and financial capital investment, consistent institutional commitment, and an ability for participation to affect change in all stages of decision making (Beierle and Cayford 2002; Charnley and Engelbert 2005; Dietz at al 2008; Leach 2006; Reed 2008; Sirianni 2009). In the years since EO 12898 was signed into law, the EPA, DOE, and DOT have experienced challenges across the range of characteristics that describe effective participatory mechanisms. In the years ahead, if federal agencies want to increase opportunities for public participation, they should consistently invest more staff time and agency resources in a range of community outreach efforts. This can help build trust and establish a framework for effective participation if environmental justice problems arise. Federal agencies face a diverse and decentralized political system. Building agile organizational capacity to partner with state and local governments will remain essential in the years ahead. Similarly, federal agencies addressing environmental justice challenges would be well served by building the technical capacity of affected communities. This provides the affected communities more opportunity to partner with government agencies when tackling environmental justice problems. Across all of these efforts to engage the public, it is also imperative to design systematic interventions to allow for clear identification of what works and why when addressing environmental challenges.

Endnotes

1. Memorandum of Understanding on Environmental Justice and Executive Order 12898 (August 2011).

2. The accomplishment reports can be found at http://www.epa.gov/compliance/environmentaljustice/resources/reports/actionplans.html

3. http://www.epa.gov/air/care/basic.htm

4. http://www.epa.gov/environmentaljustice/resources/publications/grants/ej_smgrants_emerging_tools_2nd_edition.pdf contains a full description of EJ Small Grant projects.

5. This example is drawn from the following transcript: http://www.epa.gov/environmentaljustice/multimedia/transcripts/2012-09-20-community-outreach-call.pdf

6. See the DOE website: http://energy.gov/mission

7. http://energy.gov/sites/prod/files/EJStrategy_EO12898.pdf

8. See the DOT website: http://www.dot.gov/about

9. http://www.fhwa.dot.gov/environment/environmental_justice/ej_at_dot/order_56102a

10. http://www.fhwa.dot.gov/environment/environmental_justice/ej_at_dot/dot_ej_strategy

11. http://www.fhwa.dot.gov/environment/environmental_justice/ej_at_dot/2011_implementation_report

12. http://www.fhwa.dot.gov/environment/environmental_justice/ej_at_dot/dot_ej_strategy/public_comment/index.cfm

References

Agyeman, Julian, and Briony Angus. 2003. The Role of Civic Environmentalism in the Pursuit of Sustainable Communities. *Journal of Environmental Planning and Management* 46 (3): 345–363.

Baber, Walter F., and Robert V. Bartlett. 2013. Green Political Ideas and Environmental Policy. In *The Oxford Handbook of U.S. Environmental Policy*, ed. Sheldon Kamieniecki and Michael E. Kraft, 48–66. New York: Oxford University Press.

Bardach, Eugene. 1998. *Getting Agencies to Work Together: The Practice and Theory of Managerial Craftsmanship*. Washington, DC: Brookings Institution Press.

Beierle, Thomas C. 1998. *Public Participation in Environmental Decisions: An Evaluation Framework Using Social Goals*. Washington, DC: Resources for the Future.

Beierle, Thomas C., and Jerry Cayford. 2002. *Democracy in Practice: Public Participation in Environmental Decisions*. Washington, DC: Resources for the Future.

Brehm, John, and Scott Gates. 2002. *Working, Shirking, and Sabotage: Bureaucratic Response to a Democratic Public*. Ann Arbor, MI: University of Michigan Press.

Brion, Denis J. 1988. An Essay on LULU, NIMBY, and the Problem of Distributive Justice. *Boston College Environmental Affairs Law Review* 15: 437–504.

Brulle, Robert J., and David N. Pellow. 2006. Environmental Justice: Human Health and Environmental Inequalities. *Annual Review of Public Health* 27: 103–124.

Buchanan, Sariyah S. 2010. Why Marginalized Communities Should Use Community Benefit Agreements as a Tool for Environmental Justice: Urban Renewal and Brownfield Redevelopment in Philadelphia, Pennsylvania. *Temple Journal of Science, Technology, & Environmental Law* 29: 31–52.

Bulkeley, Harriet, and Arthur P. J. Mol. 2003. Participation and Environmental Governance: Consensus, Ambivalence, and Debate. *Environmental Values* 12 (2): 143–154.

Cairns, Shannon, Jessica Greig, and Martin Wachs. 2003. *Environmental Justice & Transportation: A Citizen's Handbook.* Berkeley, CA: ITS Berkeley.

Charnley, Susan, and Bruce Engelbert. 2005. Evaluating Public Participation in Environmental Decision-Making: EPA's Superfund Community Involvement Program. *Journal of Environmental Management* 77 (3): 165–182.

Coglianese, Cary. 1997. Assessing Consensus: The Promise and Performance of Negotiated Rulemaking. *Duke Law Journal* 46 (6): 1255–1349.

Coglianese, Cary. 2006. Citizen Participation in Rulemaking: Past, Present, and Future. *Duke Law Journal* 55 (5): 943–968.

Cole, Luke W., and Shelia R. Foster. 2001. *From the Ground Up: Environmental Racism and the Rise of the Environmental Justice Movement.* New York: New York University Press.

Commission on Civil Rights (CCR). 2003. *Not In My Backyard: Executive Order 12,898 and Title VI as Tools for Achieving Environmental Justice.* http://www.usccr.gov/pubs/envjust/ej0104.pdf

Daley, Dorothy M. 2013. Public Participation, Citizen Engagement, and Environmental Decision Making. In *The Oxford Handbook of U.S. Environmental Policy,* eds. Sheldon Kamieniecki and Michael E. Kraft, 487–503. New York: Oxford University Press.

Department of Energy (DOE). 1995. *Environmental Justice Strategy: U.S. Department of Energy.* http://energy.gov/sites/prod/files/nepapub/nepa_documents/Red-Dont/G-DOE-EJ_Strategy.pdf

Department of Energy (DOE). 2009. *Environmental Justice Five-Year Implementation Plan: First Annual Progress Report.* http://energy.gov/sites/prod/files/EJ_Progress_Report.pdf

Department of Energy (DOE). 2011. *Environmental Justice Five-Year Implementation Plan: Second Annual Progress Report.* http://energy.gov/sites/prod/files/20110825%20EJ%20report%20WEB.pdf

Department of Transportation (DOT). 1995. *Department of Transportation Environmental Justice Strategy.* http://www.epa.gov/environmentaljustice/resources/publications/interagency/dot-strategy-1995.pdf

Department of Transportation (DOT). 1999. *Implementing Title VI Requirements in Metropolitan and Statewide Planning.* http://www.fhwa.dot.gov/environment/environmental_justice/facts/ej-10-7.cfm

Department of Transportation (DOT). 2012. *Department of Transportation Environmental Justice Strategy.* http://www.fhwa.dot.gov/environment/environmental_justice/ej_at_dot/dot_ej_strategy

Dietz, Thomas, Paul C. Stern, and the National Research Council. 2008. *Panel on Public Participation in Environmental Assessment and Decision Making, and National Research Council (U.S.) Committee on the Human Dimensions of Global Change. Public participation in Environmental Assessment and Decision Making.* Washington, DC: National Academies Press.

Environmental Protection Agency (EPA). 1995. *The EPA's Environmental Justice Strategy.* http://www.epa.gov/environmentaljustice/resources/policy/ej_strategy_1995.pdf

Environmental Protection Agency (EPA). 1997. *Environmental Justice: 1996 Annual Report Working Toward Solutions.* Document no. EPA/300-R-97-004.

Environmental Protection Agency (EPA). 2004. *EPA Needs to Consistently Implement the Intent of the Executive Order on Environmental Justice.* Document no. 2004-P-00007. http://www.epa.gov/oig/reports/2004/20040301-2004-P-00007. pdf

Department of Energy (DOE). 2008. *Final Site-Wide Environmental Impact Statement for Continued Operation for Los Alamos, New Mexico.* http://energy. gov/sites/prod/files/EIS-0380-FEIS-Summary-2008.pdf

Environmental Protection Agency (EPA). 2011. *Plan EJ 2014.* http://www.epa.gov/environmentaljustice/resources/policy/plan-ej-2014/plan-ej-2011-09.pdf

Environmental Protection Agency (EPA). 2013. *Plan EJ 2014 Progress Report.* Document no. EPA/300-R-13-001.

Fine, James D., and Dave Owen. 2005. Technocracy and Democracy: Conflicts between Models and Participation in Environmental Law and Planning. *Hastings Law Journal* 56 (5): 901–982.

Fletcher, Thomas H. 2003. *From Love Canal to Environmental Justice: The Politics of Hazardous Waste on the Canada-U.S. Border.* Peterborough, Canada: Broadview Press.

Foreman, Christopher H. 1998. *The Promise and Peril of Environmental Justice.* Washington, DC: Brookings Institution Press.

Furlong, Scott R. 2007. Business and the Environment: Influencing Agency Policy-making. In *Business and Environmental Policy: Corporate Interests in the American Political System*, ed. Michael E. Kraft and Sheldon Kamieniecki, 155–184. Cambridge, MA: MIT Press.

Gauna, Eileen. 1995. Federal Environmental Citizen Provisions: Obstacles and Incentives on the Road to Environmental Justice. *Ecology Law Quarterly* 22:1–87.

Government Accountability Office (GAO). 2011. *Environmental Justice: EPA Needs to Take Additional Actions to Help Ensure Effective Implementation.* Document no. GOA-12-77.

Hansell, William H., Elizabeth Hollander, and Dewitt John. 2009. *Putting Community First: A Promising Approach to Federal Collaboration for Environmental Improvement.* Washington, DC: National Academy of Public Administration.

Hernandez, Willie G. 1995. Environmental justice: Looking Beyond Executive Order No. 12,898. *UCLA Journal of Environmental Law & Policy* 14:181.

Innes, Judith E., and David E. Booher. 2004. Reframing Public Participation: Strategies for the 21st Century. *Planning Theory & Practice* 5 (4): 419–436.

Kamieniecki, Sheldon. 2006. *Corporate America and Environmental Policy: How Often Does Business Get Its Way?* Stanford, CA: Stanford University Press.

Kellogg, Wendy A., and Anjali Mathur. 2003. Environmental Justice and Information Technologies: Overcoming the Information-Access Paradox in Urban Communities. *Public Administration Review* 63 (5): 573–585.

Klyza, Christoper McGrory, and David J. Sousa. 2013. *American Environmental Policy: Beyond Gridlock.* Cambridge, MA: MIT Press.

Kraft, Michael E., and Bruce B. Clary. 1991. Citizen Participation and the NIMBY Syndrome: Public Response to Radioactive Waste Disposal. *Western Political Quarterly* 44 (2): 299–328.

Larson, Kelli L., and Denise Lach. 2008. Participants and Non-Participants of Place-Based Groups: An Assessment of Attitudes and Implications for Public Participation in Water Resource Management. *Journal of Environmental Management* 88 (4): 817–830.

Laurian, Lucie. 2007. Deliberative Planning through Citizen Advisory Boards—Five Case Studies from Military and Civilian Environmental Cleanups. *Journal of Planning Education and Research* 26 (4): 415–434.

Leach, William D. 2006. Collaborative Public Management and Democracy: Evidence from Western Watershed Partnerships. *Public Administration Review* 66 (1): 100–110.

Lubell, Mark, Paul A. Sabatier, Arnold Vedlitz, Will Focht, Zev Trachtenberg, and Marty Matlock. 2005. Conclusions and Recommendations. In *Swimming Upstream: Collaborative Approaches to Watershed Management*, ed. Paul A. Sabatier, Will Focht, Mark Lubell, Zev Trachtenberg, Arnold Vedlitz, and Marty Matlock, 261–296. Cambridge, MA: MIT Press.

Macedo, Stephen. 2005. *Democracy at Risk: How Political Choices Undermine Citizen Participation and What We Can Do about It*. Washington, DC: Brookings Institution Press.

Mazmanian, Daniel A., and Paul A. Sabatier. 1989. *Implementation and Public Policy: With a New Postscript*. Lanham, MD: University Press of America.

Mohai, Paul, David Pellow, and J. Timmons Roberts. 2009. Environmental Justice. *Annual Review of Environment and Resources* 34: 405–430.

Munton, Don, ed. 1996. *Hazardous Waste Siting and Democratic Choice*. Washington, DC: Georgetown University Press.

National Environmental Justice Advisory Council (NEJAC). 2004. *Environmental Justice and Federal Facilities: Recommendations for Improving Stakeholder Relations between Federal Facilities and Environmental Justice Communities*. http://www.epa.gov/compliance/ej/resources/publications/nejac/ffwg-final-rpt-102504.pdf

Pellow, David N., and Robert J. Brulle. 2005. *Power, Justice, and the Environment: A Critical Appraisal of the Environmental Justice Movement*. Cambridge, MA: MIT Press.

Pressman, Jeffrey L., and Aaron B. Wildavsky. 1984. *Implementation: How Great Expectations in Washington Are Dashed in Oakland: or, Why It's Amazing that Federal Programs Work At All, This Being A Saga of the Economic Development Administration as Told by Two Sympathetic Observers Who Seek to Build Morals on a Foundation of Ruined Hopes*. 3rd ed. Berkeley, CA: University of California Press.

Putnam, Robert D. 2000. *Bowling Alone: The Collapse and Revival of American Community*. New York: Simon & Schuster.

Reed, Mark S. 2008. Stakeholder Participation for Environmental Management: A Literature Review. *Biological Conservation* 141 (10): 2417–2431.

Rosenbaum, Walter A. 2002. *Environmental Politics and Policy*. 5th ed. Washington, DC: CQ Press.

Rosenstone, Steve J., and John M. Hansen. 1993. *Mobilization, Participation, and Democracy in America*. New York: Longman.

Sander, Thomas H., and Robert D. Putnam. 2010. Still Bowling Alone? The Post-9/11 Split. *Journal of Democracy* 21 (1): 9–16.

Schlozman, Kay L., Sidney Verba, and Henry E. Brady. 2012. *The Unheavenly Chorus: Unequal Political Voice and the Broken Promise of American Democracy.* Princeton, NJ: Princeton University Press.

Sirianni, Carmen. 2009. *Investing in Democracy: Engaging Citizens in Collaborative Governance.* Washington, DC: Brookings Institution Press.

Spyke, Nancy Perkins. 1999. Public Participation in Environmental Decisionmaking at the New Millennium: Structuring New Spheres of Public Influence. *Boston College Environmental Affairs Law Review* 26: 263–314.

Vajjhala, Shalini P. 2010. Building Community Capacity? Mapping the Scope and Impacts of EPA's Environmental Justice Small Grants Program. *Research in Social Problems and Public Policy* 18: 353–381.

Weber, Edward P. 2003. *Bringing Society Back In: Grassroots Ecosystem Management, Accountability, and Sustainable Communities.* Cambridge, Mass.: MIT Press.

Webler, Thomas, and Seth Tuler. 2006. Four Perspectives on Public Participation Process in Environmental Assessment and Decision Making: Combined Results from 10 Case Studies. *Policy Studies Journal: the Journal of the Policy Studies Organization* 34 (4): 699–722.

Wilson, James Q. 1989. *Bureaucracy: What Government Agencies Do and Why They Do It.* New York: Basic Books.

Yackee, Jason Webb, and Susan Webb Yackee. 2006. A Bias towards Business? Assessing Interest Group Influence on the U.S. Bureaucracy. *Journal of Politics* 68 (1): 128–139.

DISCUSSION QUESTIONS

1. Most environmental legislation calls for some form of public participation, and government agencies commonly recruit people from organized groups who have expressed interest or are active around a particular issue. Do you think the agencies' strategy to facilitate public participation is effective? Why? If not, why?

2. The U.S. EPA seeks to "work with communities through communication, partnership, research, and the public participation process" and also "help affected communities have access to information which will enable them to meaningfully participate in activities, to the end". The EPA has adopted Executive Order 12898 and the Office of Environmental Justice (OEJ)? How do the EO12898 and the creation of the OEJ affect the public participation process? In what capacity did they engender an effective public participation system?

3. Federal agencies have used public participation to address environmental justice concerns. Despite some successes, they also face significant challenges. What are they? How could federal agencies minimize the potential challenges? Do you believe building trust-based networks with decentralized political systems would play a role in mitigating the challenges?

Mushroom Packages

An Ecovative Approach in Packaging Industry

By Younsung Kim and Daniel Ruedy

Introduction

Sustainability in the packaging industry has never been more important. A $400 billion USD industry globally, packaging includes the manufacture and transport of paperboard and plastics (Ernst and Young 2013). Most conventionally designed packaging is of low cost and spends little more than a year in a linear route from producer to consumer life cycle (Hopewell et al. 2009; Niero et al. 2017). In the United States, more than 35 million tons of packaging paperboard and 160 million tons of packaging plastic are produced annually to contain, cradle, and cushion goods as they make their way through the supply chain (Geyer et al. 2017; Richtel 2016). As these packaging materials are discarded and enter the waste stream, they are disposed of in landfills, recycled, or incinerated for waste-to-energy (WTE) recovery.

Being highly visible to consumers, the packaging industry receives intense scrutiny worldwide, along the entire supply chain, from raw materials to end-of-life processes, for reducing its societal and environmental impacts (Hillier et al. 2017). The increasing concerns about the rising volume of packaging wastes are also aligned with the unproductive use of resources and materials. Both packaging producers and their primary customers—original equipment manufacturers (OEM), ecommerce and brick-and-mortar retailers, and wholesalers—face pressures to reduce materials in packaging so as to mitigate packaging's price fluctuations linked to global commodities. As such, packaging producers are forced to seek out and manage stable supply chain sources due to the instability of conventional packaging feedstocks such as wood, pulp, and petrochemicals, which would lead to reduced environmental impacts (Ernst and Young 2013).

Enhancing resource efficiency has become a conventional approach when a firm attempts to address sustainability challenges for the packaging industry (Ernst and Young 2013; Hillier et al. 2017). Noticing emerging market potentials of bio-based alternatives, some other firms have considered a more innovative approach such as developing biomaterials that may have a potential for replacing conventional, petroleum-based materials (Haneef et al. 2017; Hillier et al. 2017). Once succeeded, sustainability-promoting innovations would help firms reap greater environmental benefits and create sustainable value for shareholders (Hart 2005). However, a few firms only consider the innovative approach positively, and prior literature has rarely discussed what factors would explain why a firm could undertake innovations for sustainability. Drawing on the case of Ecovative Design, a biomaterials company headquartered in New York, USA, this study explores key factors prompting the firm to innovate their package products. Founded in 2007, Ecovative Design develops an array of environmentally friendly packaging materials by growing fibers on waste like cotton seed, wood fiber, and buckwheat hulls.

In the following sections, we first introduce the Schumpeterian creative response framework as an eco-innovation, technology-based approach to contemporary social and ecological crises, which can result in ecological modernization (York and Rosa 2003; Janicke 2008; Mason 2011). Creative response framework has been used to explain the causes and the consequences of innovation in economics and in the economy. It deviates from the continuous improvement-based approach to adopting new ideas and systems within the assumption of static economic changes (Schumpeter 1947). In the subsequent section, we examine the unsustainability of packaging products. In doing so, we underpin the current waste treatment practices of packaging materials and describe the challenges for recycling packaging products. We then explain the product innovation procedures taken by Ecovative and identify key factors of its successful practices by taking an inductive reasoning approach. Our findings indicate that technical competence combined with systems-thinking skills would prompt a firm to undertake innovative sustainability changes in its products and processes. In addition, a firm's ability to engage with governments to garner financial support and market recognition would help a firm to pursue sustainability-driven innovations for the packaging industry.

Creative Destruction for Sustainability

Global sustainability challenges present prodigious opportunities for creative responses from market disrupters. Current society-nature interactions are not sustainable in that they negatively affect both vital ecological systems and human welfare and induce irreversible, long-term damages. Achieving global sustainability then involves a sheer transition to economically revolutionary processes and routines that would be necessary to remain viable without overwhelming the society and environment. It particularly implies a structural realignment of our dominant economic development paths away from energy- and material-intensive processes relying on fossil fuels toward ecological modernization in which societies modify their institutions in order to internalize environmental impacts and ecologically transform material process and consumption processes (Mol 1995; Hajer 1995; Mason 2011).

Creative destruction was coined by Schumpeter (1947), as he noticed creative responses that brought economic developments with dynamic changes in economy. According to traditional economists, economic changes are explained by indicating specific conditioning or causal factors within a given historical development. For instance, classical economic theory predicted that an increase in population would result in a fall in per capita income, as population growth may have no other effect than that predicated by classical theory. However, this is not necessarily true in actual instances. Population growth may rather derive new developments with increasing income per capita. In other cases, a protective duty may have no other effect than to increase the price of the protected commodity and, in consequence, its output. Unlike the classical economists' prediction, however, it may also induce a complete reorganization of the protected industry which eventually results in an increase in output so great as to reduce the price below its initial level. In observation of such historical economic developments and societal changes, Schumpeter reported the neglected area of economic change, creative and energizing *reactions* to changes in "condition" that can cause novel achievements being outside of the range of existing practices (Schumpeter 1947).

As such, creative response is distinct from adaptive response. Adaptive response is the process by which an economy adapts itself to a change in its data by simply adding new resources or modestly modifying the currently existing practices. Creative response is something *beyond* the range of expanding existing practices or applying the ordinary rules. In general, the three fundamental characteristics would dictate creative response: (1) It can only be fully appreciated ex post, never being understood ex ante; (2) it shapes the whole course of subsequent events and their long-run outcomes; and (3) in terms of frequency, intensity, and success, it is influenced by the quality of the personnel available at a social and sector level and by decisions, actions, and patterns of behaviors of individuals or of groups. In this light, entrepreneurship aligned with creative response is the pivotal mechanism of inducing long-term sustained economic change in a capitalist society (Schumpeter 1947).

Creative response framework is a suitable theoretical approach to putting an emphasis on innovative responses to environmental and sustainability challenges in a society. Given the emergence and prevalence of wicked environmental problems (Lackey 2007) for the late twentieth and twenty-first centuries, different organizational responses from defiance to proactiveness have been observed (Kim and Darnall 2016). Scholars in the field of sustainability management have utilized a wide array of theoretical approaches advocated in the mainstream management sciences (Starik and Kanashiro 2013), and creative response framework has been uniquely poised to explore new ideas, experiments, or innovative sustainability practices such as biomimicry, leapfrog technology, sustainable technology, and closed loops (Braungart and McDonough 2002; Hart 2005; Larson 2000).

The social actor's energizing effects that induce new developments and exhaustive disruptions can be attributed to a different set of characteristics. Typically, sustainability responses involving modest changes for pollution involve rationalization in terms of the impact on the industry structure. That means firms' green initiatives would be flourished as long as they can yield relatively short-term benefits and appeal stakeholders that concern pollution prevention and product stewardship without restructuring the existing industry. However, the biggest leaps in performance may not be driven from firms' modest sustainability responses. Rather, firms pursuing emerging technologies, new markets, new partners, new customers not served before, and entirely new stakeholders would practice "beyond greening" initiatives that entail discontinuity (Hart 2005). For instance, the automobile industry looks nothing like it did 10–15 years ago, as companies aiming at designing sustainability mobility like Tesla pushed the industry divested from carbon-intensive, gasoline-based engine vehicles to electric ones (Jay and Gerand 2015). What makes re-invention and creative destruction possible will depend on a whole new set of skills and capabilities that have the potential of being inherently clean and sustainable (Hart 2005).

Some fundamental elements for sustainability-oriented innovation involve technical competency and stakeholder engagement for market innovation (Larson 2000; Hart 2005; Jay and Gerand 2015; Kim and Darnall 2016). Not all firms would equip with such capabilities, because the process of navigating a systems-based sustainability solution would not be appealing to most firms in the mainstream market. Some visionary companies can seize the opportunity to drive redefinition and redesign of their industries toward sustainability, and innovative upstarts can unseat established firms. The value of creative disruptive process would be reaped over the long term, as it would be able to exceed competitive gains from incremental changes in sustainability. Large incumbent firms would be less likely to shift their underlying portfolio away from what used to be its core competencies and move toward the entirely new skill sets to play a completely new, different game (Prahalad and Hart 2002; de Soto 2000; Christensen 1997).

The Need for Ecological Modernization in the Packaging Industry

Market-Oriented Sustainability Opportunity

Unsustainability of plastics. An estimated 9,150 million tons of plastics have been produced since large-scale production began around 1950, of which 30% remains in use, 9% has been recycled, and 60% is in landfills or the natural environment (Geyer et al. 2017). It is also projected that by 2050, the world will produce 28,600 million metric tons of plastic polymer resins based on observed trends (Geyer et al. 2017). The recent analyses of the global plastic packaging industry place production at 78 million tons annually at a total value of $260 billion USD (World Economic Forum et al. 2016), and it is expected to reach $490 billion by 2026 (BusinessWire 2018).

Plastic packaging does not biodegrade by design. Although exposure to sunlight is known to weaken and fragment plastic, the environmental impact of the resulting millimeter-sized fragments has not been well understood (Tudryn et al. 2018; Geyer et al. 2017). Plastic manufacturing is also dependent on nonrenewable petroleum resources for its feedstock, which indicates plastic prices fluctuating closely with the cost of petroleum (Tudryn et al. 2018). Energy consumption to sustain rising plastic demand extends past production. Transportation of plastic resins from manufacturers to downstream wholesalers also comes at a cost. The US Environmental Protection Agency (USEPA) estimates on average that plastic shipments travel 497 miles, consume 0.49 million British thermal units (BTUs) of energy, and emit 0.04 metric tons of CO_2 per short ton of plastic packaging waste (USEPA 2016a).

Approximately 50% of all plastics produced are manufactured into single-use items and packaging products among them (Hopewell et al. 2009). However, plastic recycling is increasing worldwide at 0.7% annually (Geyer et al. 2017), and the recycling practice is widely accepted to reduce lifecycle net CO_2 emissions by as much as 27% over plastic manufactured from virgin raw material feedstock (Hopewell et al. 2009). Recycling rates vary among plastic types. Polyethylene terephthalate (PET), which comprises plastic water bottles and some packaging components, is frequently recycled at a recovery rate of 19.5% by weight (USEPA 2015). However, polystyrene and its variants—expanded polystyrene (EPS) and extruded polystyrene (XPS, commonly known as Styrofoam™)—are the least recycled plastics, faring at just 0.9% recovery by weight. Polystyrene's high bulk (EPS is 95% air by weight) makes its transport costly and logistically difficult, and polystyrene's polymer structure makes it an unattractive candidate economically for primary and secondary recycling. When plastic recycling facilities are unavailable or the recovered plastic is of poor quality, waste-to-energy (WTE) incineration may be a preferred disposition option over landfilling. Yet when adjusted for the utility emissions WTE avoids, there remain net anthropogenic CO_2 emissions of 1.27 and 1.25 $MtCO_2$/ton for polystyrene and mixed plastics, respectively (USEPA 2015). Irrespective of treatment options, all plastic waste management practices (e.g., WTE incineration, recycling [including plastic fraction collection and transport to facility], landfill, etc.) have negative impacts on the environment and human health. However, it is evident that no perfect waste management solution exists, and finding the most optimal way is challenging (Rigamonti et al. 2014). In addition, on a global scale, the prospects of improving sustainability of plastics, ultimately in a hope to create a circular economy, seem to be bleak in large part due to increasing demand for consumer goods, change in consumers' lifestyle with rapid urbanization, and rising consumerism in emerging economies (De los Rios and Charnley 2017). Through a full life cycle assessment, only 5% of the value of recovered plastics is retained, costing the global economy between $80 and $120 billion USD in losses annually (World Economic Forum et al. 2016).

In response to the rigidity of the mature plastics market and its well-established supply and processing systems, plastics companies in the 1980s began blending plant starch-based polymers into environmentally friendly, theoretically biodegradable polyethylene blends (Iles and Martin 2013). When biodegradability of these bioplastics fell short of standards, they were instead remarketed as "renewably sourced," since they did replace a fraction of the petroleum-based feedstock of conventional plastics with plant-based material.

Bioplastic production is typically specialized to a single commodity feedstock, such as corn or sugarcane, and the production of which may carry its own environmental costs. Diversion of these commodities to industrial uses such as bioplastic manufacture also creates competition with stocks going to food supply, instigating complex socioeconomic policy problems. For example, North America and South America are already using 37% and 27% of domestic sugar crops for ethanol production, respectively (Golden et al. 2015). Further, though their manufacture is very similar to conventional plastics, significant infrastructure investment is required where plastics manufactured from petroleum-based feedstocks are already well-established. The wide adoption of bioplastics has failed to materialize, and prices of bioplastics remain higher than petrochemical-derived plastics (PricewaterhouseCoopers 2010), with a total production capacity of about 4 million metric tons per year (Geyer et al. 2017). Given these limitations, best-case projections predict bioplastics will at peak satisfy no more than 20% of global plastics demand (AT Kearnery 2012).

Unsustainability of paperboard. Paperboard containers travel an average of 675 miles from manufacturer to primary customer, consuming 0.67 million BTUs of energy and emitting 0.05 mT of CO_2 per short ton shipped (USEPA 2015). Though recycling is preferred to divert paperboard from landfill, paperboard that is contaminated or otherwise not suitable for recycling drives the retention rate down from an ideal 1:1 ratio (i.e., 100% of recovered paperboard is recycled) to an actual retention rate of 93.5% (USEPA 2016b). Though a paperboard product may be labeled as 100% post-consumer content, the increment not retained means long-term paperboard recycling operations will yield diminishing returns over time.

Compared to plastics, paperboard's recycling record is better at 89.5% in 2014 and, as testament to the market forces at play, up significantly from just 55% in 1993 (USEPA 2014). However, the majority of paperboard recycled is by retailers, as fewer than 50% of American consumers are estimated to utilize curbside paperboard recycling programs (Feiner 2017). This fraction of cardboard waste thus remains difficult to divert from landfill to recovery.

Regulatory Demands for the Sustainable Packaging Industry

Aside from market forces, some government policy and regulatory actions have pushed packaging producers, OEMs, retailers, and consumers to adopt more sustainable business behaviors and models.

In the European Union (EU), the Registration, Evaluation, Authorisation and Restriction of Chemicals (REACH) Regulation passed in 2006 charges private industry to determine and register the physical, environmental, and toxicological properties of substances used in quantities greater than 1 ton annually in the manufacture of goods. Suppliers and manufacturers are then required to communicate this information to downstream users, empowering OEMs and consumers to account for sustainability in their purchase decisions. As a driver for minimization, diversion, and recovery of paperboard packaging, the 1999 EU Landfill Directive called for a 50% reduction by 2009, in landfilled biodegradable waste from 1995 levels, and a 65% reduction by 2016 (PricewaterhouseCoopers 2010).

In the United States, federal regulations pertaining to the packaging industry fall under the specialized Federal Food, Drug, and Cosmetic Act but are only relevant to packaging material that comes into contact with food. Other statutes include the Toxic Substances Control Act (TSCA), the Emergency Planning and Community Right-to-Know Act (EPCRA), the Clean Air Act (CAA), the Clean Water Act (CWA), and the Resource Conservation and Recovery Act (RCRA). Though these latter statutory authorities provide for reporting requirements and adherence to some emission limits at various points in the supply chain, none expressly regulates environmental or sustainability standards for packaging. An exception is the Federal Trade Commission's (FTC's) Guides for the Use of Environmental Marketing Claims which prohibit packaging producers from making deceptive claims about compostability, degradability, recyclability, and recycled content (16 C.F.R. 260).

In terms of recycling promotion policies, half of the states have data on curbside consumer recycling programs, totaling 4,371 and serving a population 87.9 million, roughly a quarter of whom live in California (Van Haaren et al. 2010). Nationwide, average per capita municipal solid waste generation is estimated at 1.28 tons annually, of which just over a quarter is recycled (Van Haaren et al. 2010). With the void of direct federal regulation, state and local governments have taken initiatives to implement their own policies; both Washington, DC, and New York have implemented bans on EPS. Acknowledging the increasing landfill volume that packaging waste occupies and in a significant step toward meeting an ambitious 80% waste reduction goal by 2032, the Washington, DC, Department of Energy and Environment banned the use of EPS in packaging effective January 1, 2016, for food containers and January 1, 2017, for all packaging containers (DOEE 2014). Pursuant to Local Law 142 of 2013, New York City's Department of Sanitation recently determined food-service foam cannot be effectively recycled and recommended a city-wide ban starting November 12, 2017, with enforcement beginning on May 14, 2018 (DSNY 2017).

Eco-Innovation Opportunity Cumulated

In 2016 e-commerce giant Amazon delivered in excess of a billion packages (Green 2017), evidencing the tremendous success of e-commerce, but an ominous milestone from a sustainability perspective. As a whole, the e-commerce industry is valued at $350 billion and doubled in size between 2011 and 2016 (Richtel 2016). Leading online grocery markets are projected to double by 2020 (WEF 2016). Amazon for years has epitomized wasteful shipping and cardboard use—justified or not—based on its practices of packaging goods within cardboard shipping containers bearing its distinctive logo. Indeed, the successes of Amazon and ecommerce in general seem to suggest a boom in cardboard production and associated waste generation; e-commerce now accounts for 10% of all US retail (Howland 2017). But reports from the industry group Fibre Box Association (FBA) indicate that cardboard quantities shipped by US companies have actually decreased modestly since 1995 (Dove 2017).

However, a net decrease in cardboard shipped does not mean waste has been reduced or eliminated. When intrinsic costs of packaging materials are low and shipping fees are fixed, retailers and e-commerce sellers have little incentives to adopt sustainable business practices or materials into their operations and distribution. The rise of e-commerce and home delivery direct to consumers means more cardboard and plastic packaging—as well as more secondary shipping containers—can make their way directly to front stoops and business-receiving departments. The result is wasteful, producing environmentally costly over-packaging and growing waste streams.

Both consumers and shippers appear to have taken notice of this trend. Amazon's customer container feedback program has received more than 33 million responses—including comments, complaints, and photographs—from customers (Richtel 2016). In an effort to embrace an emerging positive marketing opportunity, Amazon introduced a "frustration-free packaging" option that its customers can select with their other shipping instructions (Gopaldas 2015). The option instructs Amazon's fulfillment services to forgo secondary shipping packaging to rely instead on carefully designed, minimalist OEM packaging and is credited with eliminating 83 million unnecessary cardboard boxes from shipments in 2016 alone (Pierce 2017). Further, Amazon has begun utilizing machine learning in the algorithms to determine packaging material and method (Green 2017). Informed by both cost data and customer comments, the algorithms seek to continuously optimize packaging configuration for reduced packaging and shipping costs and wastes at the consumer end (Green 2017).

Likewise, both UPS and FedEx have implemented billing systems based on dimensional weight—a measure of package volume relative to its actual weight (UPS 2017). Such measures are widely believed to incentivize OEMs and e-commerce retailers to optimize their packaging solutions to lightest and smallest possible.

While diversion of paperboard and plastic packaging waste to recycling or WTE incineration is certainly preferred over landfill disposal, many nuances influence the sustainability of this producer-to-consumer business model. Despite the good news of increased cardboard and plastic recycling or WTE rates, no matter how efficient or what material retention percent is reached, recycling at best delays waste disposal rather than eliminates it, and there remains not insignificant CO_2 emissions associated with conventional recycling and WTE incineration (Geyer et al. 2017; USEPA 2016a). Bioplastics then represents an important shift from traditional plastics manufacture, but it and other renewably sourced packaging—incrementally greener than conventional petrochemical plastics or paperboard sourced from virgin wood pulp—seem to constitute only adaptive responses to changing economic conditions (Schumpeter 1947).

Significant advances in sustainable packaging can be only made through source reduction—and ideally elimination. For example, USEPA estimates that source reduction of polystyrene from both current feedstock mixes and virgin inputs could reduce greenhouse gas (GHG) emissions by 2.5 $MtCO_2$ per short ton of source reduced (USEPA 2015). For paperboard, the reduction is more dramatic at 5.59 and 8.1 $MtCO_2$ for mixed recycled and virgin feedstock, respectively (USEPA 2015).

When upstream packaging suppliers are distanced from end users, sustainability in upstream manufacturing may not have much incentive to improve (Foerstl et al. 2015). However, as stakeholder interest in sustainability increases, as it has for Amazon, holdout firms place themselves at competitive risk by not adopting sustainable practices and may further fail to engage untapped customer bases with nascent interests in sustainably sourced materials and products (Foerstl et al. 2015). Facing pressure from expanding regulation and interest from primary and secondary customers for sustainable alternatives, the $400 billion USD packaging industry is therefore primed for disruption (Ernst and Young 2013). When assigned a monetary value, these regulatory- and altruistic-driven interests comprise a portion of a staggering $10 trillion USD in projected cumulative eco-innovation investment worldwide by 2020 (Boons et al. 2013).

Defining and Developing Sustainable Package

Sustainable Packaging Considering Business Performance and Environmental Concerns

Sustainable development and sustainability have become the focus of mainstream management studies and practices. However, the concept of sustainability has not yet been understood very clearly by corporate managers or the general public. This is in part because the concept of sustainability has been adapted to address very different challenges, ranging from the planning of sustainable cities to sustainable livelihoods, sustainable agriculture to sustainable fishing, and the efforts to develop common corporate standards in the UN Global Compact and in the World Business Council for Sustainable Development. Despite this creative ambiguity and openness to interpretation, sustainable development has evolved a core set of guiding principles and values, based on the Brundtland Commission's standard definition to meet the needs, now and in the future, for human, economic, and social development within the restraints of the life support systems of the planet (Kates et al. 2005).

As much as sustainability or sustainable development is vague in its definition, there is no clear understanding about what constitutes "sustainable packaging." A widely agreed-to and accepted understanding would be critical in the societal pursuit for promoting sustainability in the packaging domain and leading to associated business development (James et al. 2005).

As one of the semantic efforts to define sustainable packaging, Verghese and Lewis (2005) undertook a stakeholder survey in partnership with the Sustainable Packaging Alliance (SPA). The SPA's sustainable packaging definition took into consideration the role packaging plays in our social and economic systems. It also accounted for the need to meet environmental goals and reduce harm to humans and ecosystems. As such, the SPA's sustainable packaging definition includes four levels, which are society, packaging material, packaging system, and packaging component. It also identifies four different principles: *effective*, *efficient*, *cyclic*, and *safe*. The *effective* principle means that products should be packaged as they would be delivered from producers to consumers. The *efficient* principle seeks to maximize material and energy efficiency in every step of packaging, storage, transport, and handling. The *cyclic* principle is aligned with a closed-loop system, increasing recycling, reuse, and ease of disassembly and assembly. Lastly, the *safe* principle aims to minimize the human and ecological risks from packaging components, being subject to the precautionary principle. (The precautionary principle is defined by Principle 15 of the Rio Declaration. Under the precautionary principle, if there are threats of serious or irreversible damage, lack of full scientific certainty shall not be used as a reason for postponing cost-effective measures to prevent environmental degradation [United Nations 1992].) (Pielke 2002; Raffensperger and Tickner 1999). Table 2.2.1 summarizes the four-level definition of sustainable packaging. Packaging is assumed to be *sustainable* and support sustainable development if the four principles are met. While this is an earlier attempt for conceptualization, the concept well represents the multifaceted dimensions of packaging in consideration of the elements of sustainability, economy, society, and the environment (Verghese and Lewis 2005).

Table 2.2.1 SPA's sustainable packaging definition

Principle	Description	Levels at which the principle is applied
Effective	It adds real value to society by effectively containing and protecting products as they move through the supply chain and by supporting informed and responsible consumption	Society
Efficient	Packaging systems are designed to use materials and energy as efficiently as possible throughout the product life cycle. This should include material and energy efficiency in interactions with associated support systems such as storage, transport, and handling	Packaging system
Cyclic	Packaging materials are cycled continuously through natural or (industrial) technical systems, minimizing material degradation and/ or the use of upgrading additives	Packaging material
Safe	Packaging components do not pose any risks to human health or ecosystems. When in doubt, the precautionary principle applies	Packaging component

Source: Adapted from James et al. (2005)

As another well-recognized definition, sustainable packaging is also defined by the Sustainable Packaging Coalition (SPC), a project of GreenBlue which is dedicated to the sustainable use of materials in society. The SPC's sustainable packaging concept highlights a closed-loop system and promotes the five principles of (1) responsible sourcing, (2) optimization for efficiency, (3) effective recovery, (4) nontoxic, and (5) low impact. Further outlining the five principles, eight criteria were developed to define sustainable packaging (Table 2.2.2).

Table 2.2.2 SPC's definition of sustainable packaging

Criterion	Characteristic
1	Is beneficial, safe, and healthy for individuals and communities throughout its life cycle
2	Meets market criteria for performance and cost
3	Is sourced, manufactured, transported, and recycled using renewable energy
4	Optimizes the use of renewable or recycled source materials
5	Is manufactured using clean production technologies and best practices
6	Is made from materials healthy throughout the life cycle
7	Is physically designed to optimize materials and energy
8	Is effectively recovered and utilized in biological and/or industrial closed-loop cycles

Source: SPC (2011)

The eight criteria presented blend broad sustainability and industrial ecology objectives with business considerations and strategies that address the environmental concerns related to the life cycle of packaging. These criteria relate to the activities of the packaging value chain and define the areas in which we actively seek to encourage transformation, innovation, and optimization.

Indicators for Sustainable Packaging Development

Sustainable packaging development would not be straightforward even with a clear definition. Translating the definition into more specific targets or performance indictors would be useful to implement sustainable development principles in product packaging development. The proposed key performance indicators for the SPA's four sustainability packaging principles underscore two focal points, to reduce product waste and to improve functionality. Table 2.2.3 provides 21 indicators that can assist in the process of reaching a state of sustainability in packaging.

Several sustainability packaging assessment tools have been used to evaluate and compare packaging with other options. The Packaging Impact Quick Evaluation Tool (PIQET), developed by SPA, is a tool for rapid environmental impact assessment of packaging systems. This web-based software tool uses environmental indicators, based on the LCA methodology. PIQET functions as a credible, business-ready tool for multi-criteria packaging environmental decision-making and guides materials selection, packaging redesign or packaging performance, and evaluation of environmental requirements (Verghese and Lewis 2005).

Table 2.2.3 SPA's sustainable packaging indicators

Sustainable packaging principle	Sustainable packaging indicator
1. Effective	1.1 Reduces product waste 1.2 Improves functionality 1.3 Prevents overpackaging 1.4 Reduces business costs 1.5 Achieves satisfactory return on investment (ROI)
2. Efficient	2.1 Improves product/packaging ratio 2.2 Improves efficiency of logistics 2.3 Improves energy efficiency (embodied energy) 2.4 Improves materials efficiency (total amount of material used) 2.5 Improves water efficiency (embodied water) 2.6 Increases recycled content 2.7 Reduces waste to landfill
3. Cyclic	3.1 Returnable 3.2 Reusable (alternative purpose) 3.3 Recyclable (technically recyclable and system exists for collection and reprocessing) 3.4 Biodegradable
4. Safe (clean)	4.1 Reduces airborne emissions 4.2 Reduces waterborne emissions 4.3 Reduces greenhouse gas emissions 4.4 Reduces toxicity 4.5 Reduces litter impacts

Source: Lewis et al. (2007)

One of the applications of the sustainability packaging assessment tool can be found in innovating novel food packaging systems. The tool helped scholars identify optimum sustainable packaging design for food that should balance potential reductions in food loss, the ratio of the environmental impact of the food to the impact of the packaging, the handling of food waste, and the handling of packaging waste (Wikström and Williams 2010). In addition, the evaluation tool surprisingly recognized reusable plastic containers for a variety of fresh produce (Levi et al. 2011; Menesatti et al. 2012; Singh et al. 2006) as a more sustainable option when compared to commonly used corrugated paper boxes (Park et al. 2014).

Inputs and guidance from professionals in different disciplines—such as designers, engineers, technologists, marketers, and environmental managers—would inform a multidisciplinary, collaborative, and holistic approach and benefit the development process of product packaging systems.

The role of packaging technologists in the industry seems to be particularly important for sustainable packaging, as they can provide comprehensive and credible information to others within and external to the organization. This information ranges from packaging material characteristics, to packaging functionality in distribution and use, to processability in manufacturing and filling, and finally to environmental impact (SPC 2011). A significant degree of understanding and skills is needed to appropriately collect and analyze such information and to be able to present the findings to internal company decision-makers (SPC 2011).

Sustainable packaging assessment tools, a wide array of stakeholders, and technologists' information can thus be instrumental in transforming packaging system for sustainability. A closed-loop flow of packaging materials can be economically robust and provide benefits throughout its life cycle, constituting a sustainable packaging system.

In the following section, we discuss the sustainable packaging development process using a case study of Ecovative Design. The company develops an array of environmentally friendly materials that perform like plastics but are made from mushrooms. The mushroom packaging is renewable and biodegradable and can be made with crop waste brought from local farms, showing the four principles of sustainability packaging: effective, efficient, cyclic, and safe.

Case Study: Ecovative Design

Company History

Ecovative is a biotech company with the mission of designing the future of sustainable materials using Mycelium Biofabrication Platform. It took its root at Rensselaer Polytechnic Institute (RPI) in Troy, New York, created by Eben Bayer, now CEO, and Gavin McIntyre, now Chief Scientist (Zeller and Zocher 2012). In 2006, the two were classmates in Inventor's Studio, an undergraduate class instructed by RPI Professor Burt Swersey. After a semester of unsuccessful ideas, Bayer pitched the concept of a mushroom-based biopolymer as a replacement for traditional plastic insulation to Swersey. Spurred by the entrepreneurial spirit of their academic backgrounds in product design and engineering, and sharing concern for environmental responsibility, Bayer later approached McIntyre with his idea for a novel mushroom-based material to replace conventional plastics. Though the two had job offers, Swersey provided guidance and funds for the pair to pursue the idea (VentureWell 2014). Instead of following traditional career paths after graduation, the two then founded Ecovative in 2007 (Knapp 2015).

In 2008, Bayer and McIntyre travelled to Amsterdam, Netherlands, to compete in the National Postcode Lottery Green Challenge (NPCLGC) where they won the top prize of $750,000 USD for a technology to mitigate climate change (VentureWell 2014). After winning the NPCLGC, Bayer and

McIntyre transitioned their technology to the packaging industry as a cost-competitive replacement for polypropylene and polystyrene-based foams.

In the years since, Bayer, McIntyre, and their team have evaluated countless fungi species for desired ecological traits and tested their growth in various substrate media, from woodchips Bayer observed at his Vermont farm growing up to agricultural wastes of cotton burr and corn stalks.

During the stage of product development, Ecovative has experimented with shaping the material during growth by applying a mold, akin to the additive process of three-dimensional printing, or through post-growth subtractive processes, such as machining or cutting, to produce or refine bulk growth to a specified shape (Bayer et al. 2011). To augment volume and strength, agricultural waste or wood salvage may be incorporated to create a composite material (Tudryn et al. 2018).

As of 2018, Ecovative has filed for 15 patents and obtained 9 patents with the biofabrication technologies and processes to develop high-performance, sustainable materials, and products. The company's MycoFLEX platform is licensed to other manufacturers making packaging like Dell and Ikea. The platform has been also used for different consumer products that have sustainability challenges, including biofabricated leather that uses the network of mycelial fibers to create the look and texture of a hide from a cow and 3D-print artificial hearts and other body parts in the regenerative medicine industry (Peters 2018). The company employs nearly 50 employees, operating 2 warehouses, Eco-HQ in Troy, New York, at 32,000 square feet, and Eco-East in Green Island, New York, at 20,000 square feet, and generates over $1 million USD in revenue and 1 million pounds of manufactured materials annually (Ecovative Design 2018).

Ecovative's Product Innovation

The Fungi kingdom represents tremendous diversity at an estimated 1.5 million species (Hawksworth 2001). Among its other superlatives, the kingdom boasts the largest living organism on earth: a specimen of *Armillaria ostoyae* discovered in the Blue Mountains of Oregon is believed to be 2,384 acres in size and between 2,400 and 8,650 years old (Casselman 2007). Abundant fungi serve important ecological roles in the natural environment and are a source of food and medicines such as antibiotics for humans. Fungi have been growing mycelia, their unassuming chitinous root structures, for millennia.

What distinguishes Ecovative's materials is not the mycelia themselves but its leaders' visionary of harnessing fungi's natural, biological growth for niche applications ripe for disruption. As Bayer said to a TED audience in 2010, "the things that these organisms do are far more technologically advanced than anything we can dream of doing today with nanotechnology, with silicon technology, with anything we've gotten, and then by harnessing these innate properties of these organisms, these things that you would be used to seeing in your daily lives, like grass, we can do some incredible things for our planet and some really incredible things for the people living on this planet (Bayer 2011)."

In 2015, Forbes Magazine named Bayer to its eponymous "30 Under 30 List in Manufacturing" (Knapp 2015). When asked about his vision for industrial materials grown from mushrooms, Bayer recounted observing mushrooms' characteristic clumping of wood chips he would shovel for fuel while growing up on a Vermont farm (Schiffman 2013). Yet even as its materials enter the automotive, furniture, and construction markets, Ecovative maintains its original vision as an expanded polystyrene (EPS) replacement, selling its specialized Mushroom packaging in stock shapes including corner protectors and standard molded shippers. Ecovative also offers manufacture of custom packaging molds above a volume threshold (Ecovative Design 2018).

Under Bayer and McIntyre's leadership, Ecovative leveraged and received numerous government grants and contracts, including US Department of Agriculture's (USDA) Wood Innovation Grant for "Scaled Demonstration of Biological Resin System to Expand the Non-structural Engineered Wood Market," seven awards from US EPA Small Business Innovation Research (SBIR) program, and in June 2017, $9.1 million by Defense Advanced Research Projects Agency's Engineered Living Materials (ELM) program (USDA 2017; USEPA 2017; Ecovative Design 2018). Beyond funding research and development, government assistance provided Ecovative with a test bed to refine and optimize substrate blends to produce material of desired physical specifications, valuable third-party cost and performance data, and exposure to potential stakeholders and niche customers (USDA 2013; Holt et al. 2012). Among them is the National Oceanic and Atmospheric Administration's (NOAA) specialized requirement for single-use readily biodegradable launch vehicles for its tsunami buoys (USDA 2013) which are impractical to recover from the ocean once deployed from ships.

Furniture. Ecovative has recently explored processes to layer and press its materials into densities and rigidities suitable for furniture and wood flooring (Knapp 2015) and, in 2016, launched Ecovative Interiors with a limited product line. Offered directly to consumers, MycoBoard is currently used in chair backs by furniture designer Gunlocke and is also available in finished interior acoustic tiles (Ecovative Design 2018). Preliminary evaluations of Ecovative's prototype construction materials—structural insulating panels among them—found them comparable or superior to EPS (USEPA 2012).

Automotive. Bayer and McIntyre have also identified the high plastic demand of the automotive industry as a potential market and designed and tested sound-absorbing acoustic panels for automotive bodies and interiors (Pelletier et al. 2013). Results are favorable. In evaluating blends of various agricultural waste substrate, Pelletier et al. (2013) found that even Ecovative's worst-performing composite, utilizing waste cotton bur fiber as a substrate, demonstrated acoustic absorption up to 75% at 1000 Hz. Incentivized by fuel efficiency performance standards, automakers are increasingly turning to lightweight plastic components as alternatives to steel in manufacturing. Plastics contributed to just 6% of vehicle weight in 1970. In 2010, that share was up to 16% and is projected to increase to 18% by 2020 (AT Kearney 2012). Once vehicles reach the end of life, recent EU legislation requires the recycling of 60% of plastic components (AT Kearney 2012). Ecovative's solution is not only recyclable but also fully biodegradable, enabling automakers to meet and exceed increasingly stringent standards.

Construction. As one of the Ecovative's first incarnations of a viable construction material, Greensulate™ was marketed as a biodegradable alternative to conventional EPS-based insulating panels (USEPA 2009). As Ecovative scaled up its manufacturing and increased its visibility, it gained the attention of architectural designers and engineers. In 2014, architect David Benjamin installed a 40-foot-tall "Hy-Fi" pavilion at the Museum of Modern Art in New York, constructed from mycelium and corn stalk waste (Peters 2014). The design team was able to consider local sources of agricultural waste and, once the temporary installation was ready for removal, arranged for the bricks to be recycled in the Queens Community Gardens.

Though the "Hy-Fi" pavilion was perhaps more an installation art than a building prototype, it did demonstrate proof of concept and drew attention to construction applications for Ecovative's mushroom materials. Perhaps most importantly, Ecovative demonstrated the Hy-Fi's construction feedstock could be sourced from local agricultural operations. By working with communities to refine its supply chains in favor of local sources, Ecovative and its partners could tap into nascent consumer preferences for and pioneer a locally sourced construction materials market, similar to those that popularized now ubiquitous locally grown produce and farm-to-table restaurants.

Discussion

Key Driving Forces of Ecovative's Success

Unsustainable practices in the packaging industry are not considered new due to petroleum-based plastic's dominance, but their social and environmental costs have not been fully internalized (Ernst and Young 2013; Hillier et al. 2017). However, the rising volume of packaging plastics along with increasing consumption of paperboard packaging explains why diverse stakeholder groups including consumers have raised serious concerns over the unsustainability of packaging materials through the supply chain (Geyer et al. 2017; Richtel 2016). Some firms in the packaging industry have thus acted to lower social and environmental impacts of packaging, although most responses lead modest changes within the existing industrial practices. Common are projects related to resource efficiency improvement, recycling, or waste-to-energy recovery. Seen in this light, Ecovative's response involving the entrepreneurial procedures of fabricated biomaterials from mycelia is creative and disruptive. It can shape the contour of packaging industry, creating situations from which there is no bridge to those situations that might have emerged in its absence. Ecovative's creative response leaves us with a question of what drives the firm's creative response. This inductive case study pivots the firm's unique capabilities, technical competence for systems thinking, and the ability of engaging the stakeholder matrix. Each element is investigated in detail below.

Technical competence for systems thinking. Ecovative has harnessed the power of mushroom to create natural, biodegradable packaging and materials for potential use in multiple industries. Rather than produce, manage, and ideally minimize its own waste stream, Ecovative has focused on the development of technology which is capable of drawing on the waste streams from agricultural processes (Holt et al. 2012). The company's technological innovation is radical and unique in that it can horizontally integrate agricultural organizations, itself, and primary and secondary customers in a novel, innovative way (Larson 2000).

Ecovative's ability to disrupt existing practices employed by incumbents originates from the founding team's profound understanding about natural ecosystem and organisms, competitive imagination, and entrepreneurial leadership. Bayer touts the opportunities of organisms' innate properties and biological processes, many of which cannot be duplicated by technology: "with biology you can tell these organisms to do extra things; to make a compound while you're growing. That's our long-range vision for biofabrication" (Knapp 2015). He also held entrepreneurial leadership, through which he successfully organized a core group of people willing to take risks of failures and to take advantage of a sustainability-driven innovation opportunity that is not quite regarded as appealing.

Because packaging is a low-value good, it is generally not profitable to transport to long distances to primary or secondary customers. To an extent, this characteristic has protected packaging producers in developed countries from lower prices of overseas competitors (Ernst and Young 2013) and insulated established, geographic regions in the United States from innovators like Ecovative, were it to follow a conventional supply chain model.

Ecovative's packaging solution instead upends the traditional manufacturing packaging supply models and connects various nodes into a novel, cross-sector supply chain. Its packaging's primary raw material, agricultural waste, is readily available nationwide. Ecovative's packaging products do not require the capital and infrastructure—not to mention energy-intensive manufacturing and associated waste streams—of traditional cardboard or even bioplastics. Should they be so emboldened, retailers can eliminate the need for a packaging manufacturer, transporters, as well as associated emissions and waste at each node, by producing their own packaging materials in-house and supplanting distant suppliers of paperboard pulp or plastic pellets with a local provider of agricultural waste feedstock (Verghese and Lewis 2007). With little infrastructure investment or utility costs, an OEM could conceivably source its agricultural waste locally within tens of miles (rather than the hundreds of miles characteristic of both paperboard and plastic packaging) and produce its own customized packaging in-house under license from Ecovative. Bayer recognizes supply chain economics are just as important as product performance: "on the raw materials side, we've been able to drive the raw material costing down to be at price parity with the styrene and polyethylene that's used in packaging so we're just finally getting to realize that dream of a triple-threat value proposition" (VentureWell 2014).

Engaging the stakeholder matrix. Though Ecovative's applications of fungi mycelium for materials are radical, the fundamental problem of unsustainability in packaging it sought to address was not.

Independent of their lifecycle environmental performance or adoption by industry, high post-consumer content materials and bioplastics likely lowered the acceptance threshold of biomaterials for Ecovative's future customers. With exception of stakeholders already entrenched in the industry, Ecovative's stakeholders included government and industry incumbents, which is very unique to new entrants to the industry (Hall and Martin 2005; USEPA 2009). Bayer is very familiar with the importance of engaging the stakeholder network: "We unite a lot of stakeholders. Commercially, we work with really big companies either on direct customer access or on future projects, and we also work with the government. We've sort of connected to as many stakeholders in our space as we can" (VentureWell 2014).

Ecovative's interactions with government entities crossed both primary (as bona fide or potential customers of Ecovative's bio-based labeled products) and secondary stakeholder (as influencers of regulatory action) domains. Driven by Executive Order 13693, Planning for Federal Sustainability in the Next Decade, the US Department of Agriculture's (USDA) BioPreferred program sets minimum standards for percent bio-based content among more than 97 product categories (e. g., cleaners, carpet, paint, etc.) and mandates their purchase by federal agencies and contractors (USDA 2015b; Golden et al. 2015). With federal consumption of goods and services valued at $445 billion USD annually (Golden et al. 2015), through USDA's BioPreferred certification, Ecovative enjoys a competitive advantage to a somewhat captive primary stakeholder customer (USDA 2015a, b). Ecovative's choice to participate in USDA's Agricultural Research Service (ARS) and EPA's Small Business Innovative Research (SBIR) programs engaged these agencies as secondary stakeholders and valuable co-producers, providing capital, lending technical support, and helping to establish objective third-party performance data to assess commercial viability.

Being equally important to collaborating with governmental agencies, paramount to navigating growth is carefully selecting supply chain partners and understanding the requirements of primary and secondary customers (Golden et al. 2015). Ecovative's unique supply chain links together diverse stakeholders in novel relationships. Ecovative entered into an existing packaging industry market by large, established firms. Among them was Sealed Air, a $7 billion company and purveyor of packaging staples like Bubble Wrap', and a customer base that includes distribution giants US Foods and Kroger Corporation (Sealed Air 2012). An unlikely ally, Sealed Air, has become a manufacturing partner to Ecovative in operations of its Eco-East facility.

Conclusion

The unsustainability of packaging industry has been under high scrutiny from various stakeholders (Ernst and Young 2013). Being highly visible to consumer, the packaging industry is thus compelled to decarbonize and dematerialize the industry across the entire supply chain, from raw materials to end-of-life processes for reducing its societal and environmental impacts (Hillier et al. 2017).

Identifying the need to modify their production practices and processes, some firms in the packaging industry have attempted to seize a market opportunity for environmentally friendly packages and bio-based materials that can replace petroleum-based ones.

More than a decade ago, Ecovative Design, a biomaterials company growing packaging products with mycelium, envisioned the need of responsible product designs in packaging. Rather than incrementally changing the status quo of conventional packaging products, the company has brought an entirely new material solution to market by harnessing the natural mycelia root structures of fungi. Ecovative is then able to produce materials comparable in cost and performance to EPS and engineered cardboard securely bound by pressure physically and enzymatically by a completely natural, chitinous biopolymer (Haneef et al. 2017). The company enters the automotive, furniture, and construction markets, illustrating boundless product application options that have not been yet investigated or claimed (Ecovative Design 2018). Companies partnering with Ecovative for sustainable packaging alternatives include large retailers like Dell and Ikea. Innovative firms seeking bio-based materials (e.g., biofabricated leather) and growing meat without livestock are collaborating with Ecovative to find out the optimal applications of the Ecovative's innovative MycoFLEX platform. The medical start-ups focusing on regenerative medicines also plan to use the Ecovative's approach in an attempt to develop 3D-print artificial hearts and other body parts (Peters 2018).

Relying on the inductive case study approach, this study finds that Ecovative's creative innovation would be attributed to technical competence for systems thinking and the firm's capacity of engaging the stakeholder matrix. The two elements facilitate getting closed-loop thinking be transformed into the technological process of growing trays of mycelia within controlled temperature, humidity, airflow, and other factors and by which the shape and density of mycelia can be fully controlled. Utilizing the Ecovative's approach for biofabrication technologies presents an immense market prospect, because biofabrication has been emerged as a twenty-first-century manufacturing paradigm, but its development has been in its infancy (Mironov et al. 2009). A huge innovation opportunity for sustainability thus relates to sustained firm value that is not achievable solely through continuous, incremental improvement (Hart 2005).

In sum, the Ecovative's approach pivots on the significance of entrepreneurial firms' innovative initiatives based on their eco-innovation capabilities, technical competence for systems thinking, and ability of engaging with diverse stakeholders. The approach has proven enough to disrupt the ruts of existing practices in packaging industry. Extending this approach and encouraging firms' creative response for discontinuity, society can advance its sustainability and radically lower social and environmental impacts from today's old, existing unsustainable practices.

Exercises in Practice

- Identify a sustainability challenge in an industry that requires a systems-thinking approach to innovate the entire industry's unsustainable practices. Discover which kind of technical competence is needed to stymie the traditional production patterns across the unsustainable supply chain.
- Explore available opportunities to engage with stakeholders that may provide essential financial support and technical advice during the process of ecological modernization and creative disruption for sustainability.

Key Lesson for Engaged Sustainability

This case study illustrates the potential of technical competence along with a systems-thinking skill could upend the highly mature industrial supply chain. Engaging with a diverse set of stakeholders would also be another element to break through the unsustainable pattern of industry practices.

Reflection Questions

- How can new market entrants seize the opportunity to innovate a highly mature industry that could be rigid in bringing new innovative changes and ecological modernization?
- How can an innovative firm balance direct stakeholder engagement while protecting its intellectual property?
- What other industries could follow Ecovative's licensing model to get their sustainability ideas to market?
- What challenges may be presented if new market entrants target an incumbent solution and its associated supply chain?
- How can regulators not only incentivize firms with technical competence but also promote the scalability of the sustainability solutions to enhance industry-wide transformation for sustainability?
- Are there any examples that are synonymous with Ecovative in other industries? If then, what are the features that would make Ecovative distinct from those illustrations?

References

AT Kearney. (2012). Plastics. The future for automakers and chemical companies. https://www.atkearney .com/documents/10192/244963/Plastics-The_Future_for_Automakers_and_Chemical_Companies .pdf/28dcce52-affb-4c0b-9713-a2a57b9d753e

Bayer, E. (2011). Eben Bayer: Drinking trees [Video file]. Retrieved from https://www.youtube.com/watch?v=VTsH8qgIb80

Boons, F., Montalvo, C., Quist, J., & Wagner, M. (2013). Sustainable innovation, business models and economic performance: An overview. *Journal of Cleaner Production, 45*, 1–8.

Braungart, M., & McDonough, W. (2002). *Cradle to cradle. Remaking the way we make things.* New York: North Point Press.

BusinessWire (2018). The global market for plastic packaging (2017–2026) to grow at 5.4% CAGR-environmental concern is hampering growth. https://www.businesswire.com/news/home/20180828005731/en/Global-Market-Plastic-Packaging-2017-2026-Grow-5.4

Casselman, A. (2007, October 4). Strange but true: The largest organism on earth is a fungus. *Scientific American.* https://www.scientificamerican.com/article/strange-but-true-largest-organism-is-fungus/

Christensen, C. (1997). *The innovator's dilemma.* Boston: Harvard Business School Press.

De los Rios, I. C., & Charnley, F. J. S. (2017). Skills and capabilities for a sustainable and circular economy: The changing role of design. *Journal of Cleaner Production, 160*, 109–122.

De Soto, H. (2000). *The mystery of capital.* New York: Basic Books.

Department of Energy & Environment (DOEE). (2014). Sustainable DC omnibus amendment act of 2014. https://doee.dc.gov/foodserviceware

Department of Sanitation, New York City (DSNY). (2017). Determination on the recyclability of food-service foam pursuant to local law 142 of 2013. https://www1.nyc.gov/assets/dsny/docs/2017-05-12FoamDetermination_FINAL.pdf

Dove, L. (2017). Has online shopping changed how much cardboard we use? https://science.howstuffworks.com/environmental/green-science/online-shopping-cardboard-consumption-industry-amazon.htm

Ecovative Design. (2018). How it works. https://ecovativedesign.com/how-it-works

Ernst & Young. (2013). Unwrapping the packaging industry: Seven factors for success. http://www.ey.com/Publication/vwLUAssets/Unwrapping_the_packaging_industry_%E2%80%93_seven_factors_for_success/$FILE/EY_Unwrapping_the_packaging_industry_-_seven_success_factors.pdf

Feiner, L. (2017, August 12). Why online shopping may not save the cardboard box. *Boston Globe.* https://www.bostonglobe.com/business/2017/08/11/why-online-shopping-may-not-save-cardboard-box/s2euhXKXpRhDWoOzmu2CNO/story.html

Foerstl, K., Azadegan, A., Leppelt, T. L., & Hartman, E. (2015). Drivers of supplier sustainability: Moving beyond compliance to commitment. *Journal of Supply Chain Management, 51*(1), 67–92.

Geyer, R., Jambeck, J. R., & Law, K. L. (2017). Production, use, and fate of all plastics ever made. *Science Advances, 3*(7), 1–5.

Golden, J. S., Handfield, R. B., Daystar, J., & McConnell, T. E. (2015). An economic impact analysis of the US biobased products industry: A report to the Congress of the United States of America. *Industrial Biotechnology, 11*(4), 201–209.

Gopaldas, A. (2015). Creating firm, customer, and societal value: Toward a theory of positive marketing. *Journal of Business Research, 68*, 2446–2451.

Green, D. (2017). Amazon fixed the most annoying thing about receiving online orders. *Business Insider.* http://www.businessinsider.com/amazon-fixes-packaging-to-be-more-efficient-2017-12

Hajer, M. A. (1995). *The politics of environmental discourse: Ecological modernization and the policy process.* Oxford, UK: Oxford University Press. ISBN 0-19-82769-8.

Hall, J. K., & Martin, M. J. C. (2005). Disruptive technologies, stakeholders, and the innovation value-added chain: A framework for evaluating radical technology development. *Research & Development Management, 35*(3), 273–284.

Haneef, M., Cesuracciu, L., Cahale, C., Bayer, I. S., Heredia-Guerroro, J. A., & Athanassiou, A. (2017). Advanced materials from fungal mycelium: Fabrication and tuning of physical properties. *Nature, 7,* 41292. https://doi.org/10.1038/srep41292

Hart, S. L. (2005). Innovation, creative destruction and sustainability. *Research-Technology Management, 48*(5), 21.

Hawksworth, D. L. (2001). The magnitude of fungal diversity: The 1.5 species estimate revisited. *Mycological Research, 105*(12), 1422–1432.

Hillier, D., Comfort, D., & Jones, P. (2017). The packaging industry and sustainability. *Athens Journal of Business and Economics, 3*(4), 405–426. ISSN 2241-794X.

Holt, G. A., Mcintyre, G., Flagg, D., Bayer, E., Wanjura, J. D., & Pelletier, M. G. (2012). Fungal mycelium and cotton plant materials in the manufacture of biodegradable molded packaging material: Evaluation study of select blends of cotton byproducts. *Journal of Biobased Materials and Bioenergy, 6*(4), 431–439.

Hopewell, J., Dvorak, R., & Kosior, E. (2009). Plastics recycling: Challenges and opportunities. *Philosophical Transactions of the Royal Society of London. Series B, Biological Sciences, 364*(1526), 2115–2126.

Howland, D. (2017). Amazon is scrambling to minimize packaging. *Retail Dive.* https://www.retaildive.com/news/amazon-is-scrambling-to-minimize-packaging/513634/

Iles, A., & Martin, A. N. (2013). Expanding bioplastics production: Sustainable business innovation in the chemical industry. *Journal of Cleaner Production, 45*, 38–49.

James, K., Fitzpatrick, L., Lewis, H., & Sonneveld, K. (2005). Sustainable packaging system development. In W. Leal Filho (Ed.), *Handbook of sustainability research.* Frankfurt: Peter Lang Scientific Publishing.

Janicke, M. (2008). Ecological modernization: New perspectives. *Journal of Cleaner Production, 16*(5), 557–565.

Jay, J., & Gerand, M. (2015). Accelerating the theory and practice of sustainability-oriented innovation. MIT Sloan School working paper 5148-15.

Kates, R., Parris, T. M., & Leiserowitz, A. A. (2005). What is sustainable development?: Goals, indicators, values, and practice. *Environment, 47*(3), 9–21.

Kim, Y., & Darnall, N. (2016). Business as a collaborative partner: Understanding firms' sociopolitical support for policy formation. *Public Administration Review, 76*(2), 326–337.

Knapp, A. (2015, May 6). This entrepreneur is literally growing the future of manufacturing. *Forbes.* https://www.forbes.com/sites/alexknapp/2015/05/06/this-entrepreneur-is-literally-growing-the-future-of-manufacturing/#2abbc7f9210b

Lackey, R. (2007). Science, scientists, and policy advocacy. *Conservation Biology, 21*(1), 12–17.

Larson, A. L. (2000). Sustainable innovation through an entrepreneurship lens. *Business Strategy and the Environment, 9,* 304–317.

Levi, M., Cortesi, S., Vezzoli, C., & Salvia, G. (2011). A comparative life cycle assessment of disposable and reusable packaging for the distribution of Italian fruit and vegetables. *Packaging Technology and Science, 24,* 387–400.

Lewis, H., Fitzpatrick, L., Verghese, K., Sonneveld, K., & Jordon, R. (2007). Sustainable packaging redefined. http://nbis.org/nbisresources/packaging/sustainable_packaging_guidelines.pdf

Mason, M. (2011). The sustainability challenge. In J. Brady, A. Ebbage, & R. Lunn (Eds.), *Environmental management in organizations* (pp. 525–532). London: Earthscan.

Menesatti, P., Canali, E., Sperandio, G., Burchi, G., Devlin, G., & Costa, C. (2012). Cost and waste comparison of reusable and disposable shipping containers for cut flowers. *Packaging Technology and Science, 25,* 203–215.

Mironov, V., Trusk, T., Kasyanov, V., Little, S., Swaja, R., & Markwald, R. (2009). Biofabrication: A 21st century manufacturing paradigm. *Biofabrication, 1*(2), 1–16. https://doi.org/10.1088/1758-5082/1/2/022001.

Mol, A. P. J. (1995). *The refinement of production: Ecological modernization theory and the chemical industry.* Utrecht: Van Arkel.

Niero, M., Hauschild, M. Z., Hoffmeyer, S. B., & Olsen, S. I. (2017). Combining eco-efficiency and eco-effectiveness for continuous loop beverage systems. *Journal of Industrial Ecology, 21*(3), 742–753.

Park, S., Lee D. S., & Han J. H. (2014). Eco-design for food packaging innovations. In J. H. Han (Ed.), *Innovation in food packaging* (pp. 537–547). Academic. https://doi.org/10.1016/C2011-0-06876-X.

Pelletier, M. G., Holt, G. A., Wanjura, J. D., Bayer, E., & McIntyre, G. (2013). An evaluation study of mycelium based acoustic absorbers grown in agricultural by-product substrates. *Industrial Crops and Products, 51,* 480–485.

Peters, T. (2014). Sustaining the local: An alternative approach to sustainable design. *Architectural Design, 85,* 136–141.

Peters, A. (2018). Can mushrooms be the platform we build the future on? *FastCompany.* https://www.fastcompany.com/90246740/can-mushrooms-be-the-platform-we-build-the-future-on

Pielke, R. J. (2002). Better safe than sorry. *Nature, 419*(6906), 433–434.

Pierce, L. M. (2017). Amazon on creating commerce packaging that's great for all: Customers, companies and the environment. *Packaging Digest.* http://www.packagingdigest.com/optimization/amazon-on-creating-ecommerce-packaging-thats-great-for-customers-companies-and-envi ronment-2017-04-14

Prahalad, C. K., & Hart, S. L. (2002). The fortune at the bottom of the pyramid. *Strategy+Business, 26,* 54–67.

PricewaterhouseCoopers. (2010). Sustainable packaging: Threat or opportunity? https://www.pwc.com/gx/en/forest-paper-packaging/pdf/sustainable-packaging-threat-opportunity.pdf

Raffensperger, C., & Tickner, J. (1999). *Protecting public health and the environment: Implementing the precautionary principle.* Washington, DC: Island Press.

Richtel, M. (2016, February 16). E-Commerce: Convenience built on a mountain of cardboard. *The New York Times.* https://www.nytimes.com/2016/02/16/science/recycling-cardboard-onlineshopping-environment.html

Rigamonti, L., Grosso, M., Moller, J., Sanchez, V. M., Magnani, S., & Christensen, T. H. (2014). Environmental evaluation of plastic waste management scenarios. *Resources, Conservation and Recycling, 85,* 42–53.

Schiffman, R. (2013). One minute with … Eben Bayer. *New Scientist, 218*(2921), 29.

Schumpeter, J. A. (1947). The creative response in economic history. *Journal of Economic History, 7*(2), 149–159.

Sealed Air. (2012). Sealed air and Ecovative complete agreement to accelerate commercialization of new sustainable packaging material. http://ir.sealedair.com/phoenix.zhtml?c=104693&p=irolnewsArticle_Print&ID=1706817

Singh, S. P., Chonhenchob, V., & Singh, J. (2006). Life cycle inventory and analysis of reusable plastic containers and display-ready corrugated containers used for packaging fresh fruits and vegetables. *Packaging Technology and Science, 19,* 279–293.

Starik, M., & Kanashiro, P. (2013). Toward a theory of sustainability management: Uncovering and integrating the nearly obvious. *Organization and Environment, 26*(1), 7–30.

Sustainable Packing Coalition. (2011). Definition of sustainable packaging. https://sustainablepackaging.org/wp-content/uploads/2017/09/Definition-of-Sustainable-Packaging.pdf

Tudryn, G. J., Smith, L., Freitag, J., Bucinell, R., & Schadler, L. (2018). Processing and morphology impacts on mechanical properties of fungal-based biopolymer composites. *Journal of Polymers and the Environment, 26*(4), 1473–1483.

United Nations. (1992). *Rio declaration on environment and development.* Rio de Janeiro: United Nations. http://www.unep.org/Documents.Multilingual/Default.asp?DocumentID=78&ArticleID=1163

United Parcel Service (UPS). (2017). Packaging guidelines: Billable weight. https://www.ups.com/us/en/help-center/packaging-and-supplies/determine-billable-weight.page

United States Department of Agriculture (USDA). (2013). Biodegradable packaging from cotton waste. *Agricultural Research, 61*(10), 16–18.

United States Environmental Protection Agency (USEPA). (2009). *Ecovative design: Growing America's green economy with research and innovation.* Office of Research and Development. https://archive.epa.gov/ncer/publications/web/pdf/ncse-greensulate.pdf

United States Environmental Protection Agency (USEPA). (2012). Development and demonstration of a low embodied energy, construction material that replaces expanded polystyrene and other synthetic materials. EPA contract number EPD10058.

United States Environmental Protection Agency (USEPA). (2017). *Ecovative design: Greensulate| growing America's green economy with research and innovation.* Office of Research and Development. https://archive .epa.gov/ncer/publications/web/pdf/ncse-greensulate.pdf

USDA. (2015a). Driving the bioeconomy: Economic impact of the biobased products industry. Presentation by Marie Wheat, Industry Economist, USDA BioPreferred Program, to the BIO World Congress on Industrial Biotechnology. 22 July 2015.

USDA. (2015b). *Development of scaled manufacturing for mycological soilless growth media.* National Institute of Food and Agriculture. Grant number 2015-33610-23814.

USDA. (2017). U.S. Forest Service awards grants to expand and accelerate wood energy and wood products markets in 19 States. U.S. Forest Service Press Release. 24 May 2017.

USEPA. (2014). Advancing sustainable materials management: Facts and figures. https://www.epa.gov/smm/ advancing-sustainable-materials-management-facts-and-figures

USEPA. (2015). Documentation for Greenhouse Gas Emission and Energy Factors Used in the Waste Reduction Model (WARM). Office of Resource Conservation and Recovery. https://archive.epa.gov/epawaste/conserve/ tools/warm/pdfs/WARM_Documentation.pdf

USEPA. (2016a). Documentation for greenhouse gas emission and energy factors used in the waste reduction model (WARM). Office of Resource Conservation and Recovery. https://19january2017snapshot.epa.gov/ sites/production/files/2016-03/documents/warm_v14_containers_packaging_non-durable_goods_materials .pdf

USEPA. (2016b). Advancing sustainable materials management: 2016 Recycling Economic Information (REI) report. October 2016. EPA530-R-17-002.

Van Haaren, R., Themelis, N., & Goldstein, N. (2010). The state of garbage in America. *Biocycle, 51*(10), 16–21.

VentureWell. (2014). VentureWell idea to impact, *Ecovative.* https://www.youtube.com/watch?v=HqmCSml15jU

Verghese, K., & Lewis, H. (2005). Sustainable packaging: How do we define and measure it. Presented at the 22nd International Association of Packaging Research Institutes Symposium.

Verghese, K., & Lewis, H. (2007). Environmental innovation in industrial packaging: A supply chain approach. *International Journal of Production Research, 45*(18–19), 4381–4401.

Wikström, F., & Williams, H. (2010). Potential environmental gains from reducing food losses through development of new packaging—A life-cycle model. *Packaging Technology and Science, 23*, 403–411.

World Economic Forum, Ellen MacArthur Foundation and McKinsey & Company. (2016). The new plastics economy: Rethinking the future of plastics. http://www.ellenmacarthurfoundation.org/publications

York, R., & Rosa, E. A. (2003). Key challenges to ecological modernization theory. *Organization and Environment, 16*(3), 273–288.

Zeller, P., & Zocher, D. (2012). Ecovative's breakthrough materials. *Fungi, 5*(1), 51–56.

DISCUSSION QUESTIONS

1. Packaging industry faces sustainability challenges. What are the elements of unsustainability in plastics and paperboard?

2. What are the regulatory contexts of the packaging industry in the United States and in the EU? How do they present eco-innovation opportunities to the packaging industry?

3. Ecovative Design's mushroom-based packages present a possibility for improving total sustainability in the packaging industry. Using the four sustainable packaging criteria (effective, efficient, cyclic, safe, in Table 1), explain the extent to which the Evocative Design products are sustainable.

4. Creative destruction framework has been useful in understanding the disjointed, dynamic social and economic changes. Beyond the packaging industry, where are other instances where you see creative destruction phenomena bringing sustainability innovation?

The Inclusive City

Urban Planning for Diversity and Social Cohesion

By Franziska Schreiber and Alexander Carius

....

Every week, about 3 million people move to cities worldwide. Over the coming decades, such migration will contribute to an increase in the urban share of the global population from 54 percent in 2014 to 66 percent in 2050. Although migration is not a new phenomenon, the current pace of rural-urban migration, both within and between countries, is unprecedented. In developing and emerging economies, this has led to the mushrooming of megacities such as Cairo, Jakarta, Lagos, Manila, and Mumbai. However, cities are not only growing in population, but also becoming increasingly diverse and ethnically heterogeneous. This twofold process poses great challenges, as cities have to manage the multi-faceted integration of their arriving newcomers into society and urban life, as well as ensure continued social cohesion.[1]

Strong integration policies are needed that support urban migrants in finding jobs, living in socially mixed neighborhoods, learning the language, and enabling their children to go to school. In addition to policies related to education, health care, the job market, housing, and finance, the ways that cities are designed and constructed are important elements of integration policy. For example, well-designed urban patterns and functioning public spaces that serve as meeting places for urban dwellers can aid in facilitating interaction, connectivity, and social mixing—all important aspects of cohesive cities.

Various urban planning and design measures can be used to strengthen the relationship between space and social integration, helping to address the challenges that cities face with respect to migration, segregation, and socioeconomic polarization. At the national level, programs and frameworks can enable actions in cities and neighborhoods to improve the social and economic conditions of residents, as examples from Germany, Denmark, India, and South Africa illustrate. At the local level, city-wide and neighborhood planning can develop compact, well-connected and integrated urban patterns that facilitate social interaction and integration, as illustrated by case studies from Berlin, Germany; Guangzhou, China; Medellín, Colombia; and Oslo, Norway. Planning and design approaches that support "inclusive cities" and greater social cohesion include land-use planning, integrated land-use and transport planning, upgrading street networks, and public-space design.

Tackling Growing Urban Challenges

In our increasingly urbanized world, cities function as a melting pot for people with differing cultural backgrounds, religions, interests, and social status. In this context, cities and municipalities face the twin challenges of not only absorbing the influx of people from diverse social and ethnic backgrounds, but also counteracting the trend of rising socioeconomic polarization and the segregation of cities into privileged and disadvantaged neighborhoods.

These two challenges are often intertwined and need to be approached holistically. Although research indicates that "no intrinsic link between deprivation and ethnic heterogeneity" exists, there is ample evidence that poorly managed urban migration results in the marginalization and segregation of people with different backgrounds. Questions related to the impact of immigration and ethnic diversity on the social fabric in cities are being debated in countries across the globe. Such discussions have been particularly prominent in the context of the refugee crisis in the European Union, where hundreds of thousands of refugees from conflict-torn and fragile regions, such as Afghanistan, Eritrea, Iran, Iraq, Syria, and the Western Balkans, are seeking asylum.[2]

Although many cities and municipalities are demonstrating courage, flexibility, and creativity in organizing ad hoc accommodation, care, and food for new migrants, the long-term challenge will be to ensure their full integration into society and to create acceptance among the local population. The latter is related to the rise of xenophobia and to fears about the consequences of uncontrolled, overwhelming migration, such as added competition in the labor market or a decline of social cohesion. Such fears have arisen in many European countries in response to the influx of refugees, and local governments need to take these concerns seriously in order to counteract the prevailing perception of migration as a "problem."

In addition to managing the integration of immigrants, cities and municipalities must provide sufficient infrastructure to accommodate their growing and diversifying populations and to avoid the emergence of new inequalities in urban areas while fostering social cohesion. For example, local governments must meet the increasing demand for housing and provide sufficient infrastructure and basic services, such as electricity, water, sanitation, health care, and education. Cities in developing and emerging countries, in particular, often lack the capacity to meet these needs and are confronted with the sprawl of informal settlements and slums (and thus an intensification of social and spatial segregation). Between 1990 and 2012, the share of the urban population living in slum areas in developing regions increased from 35 percent to 46 percent.[3]

The huge demand for housing is a challenge in developed countries as well, where rental prices are rising rapidly and the amount of social housing is declining, with adverse impacts on the social structure in neighborhoods. According to a government-conducted housing survey, the social housing stock in the United Kingdom has declined from 5.5 million homes in 1980/81 to 3.8 million homes in 2010/11, suggesting that people increasingly face difficulties in accessing adequate, affordable, and secure housing. Although the United Kingdom was once a forerunner in providing public housing, this achievement has been undermined by recent polices, such as the "Right to Buy," under which millions of social-housing units were sold. As waiting lists for social housing lengthen due to the slow construction of new houses, not even half of the demand for this housing is being met, and the degree of spatial segregation between the rich and poor in U.K. cities is increasing.[4]

Policies are needed at the national and local levels to support integration and to counteract segregation through infrastructure measures. However, the work of urban planners and designers also can contribute greatly to social cohesion. Even though the reorganization of space to create more-integrated urban patterns (for example, socially and functionally mixed areas that are well connected and easily accessible) and physical interventions (such as urban design measures in public spaces) cannot solve the roots of social and economic problems, they can aid in creating more-inclusive cities. Karin Peters and her colleagues at Wageningen University in the Netherlands argue that "interactions in daily life between people across ethnic divides are one way of creating social cohesion, because they provide the basis for bonds between individuals." It therefore is important to consider what (and how) planning and design measures at different scales, including the national, city, and neighborhood levels, can foster social interaction and integration in social networks.[5]

Some cities and countries have successfully implemented inclusive national and local plans, policies, and measures that provided a "spatial fix" to social problems and initiated positive locational dynamics. In Colombia, the city of Medellín implemented an innovative public transport system to connect poor and formerly inaccessible districts with the rest of the city, helping to enhance quality of life, attract tourists, and reduce the level of crime in these areas; however, this move did not solve the fundamental roots of poverty of many residents. The International Organization for Migration notes that, to achieve the greatest impact, "effective national and international instruments and institutions also need to be put in place." Planning and design measures should be embedded into a broader urban-cohesion policy, which involves a range of policy approaches in the areas of education, health care, employment, housing, and finance.[6]

From Exclusion to Interaction to Cohesion

There is a common perception that the quality of public and civic life is in alarming decline worldwide. Since the 1970s, economic inequality has grown, resulting in socioeconomic polarization and spatial segregation, especially in urban areas. More than two-thirds of the urban population lives in cities where the income gap has widened sharply in the past three decades. The level of income inequality in these cities often surpasses the United Nations alert line of 0.4, based on the so-called Gini coefficient, which ranges from 0 (everyone has the same income) to 1.0 (maximum inequality of income). (See Figure 2.3.1.)[7]

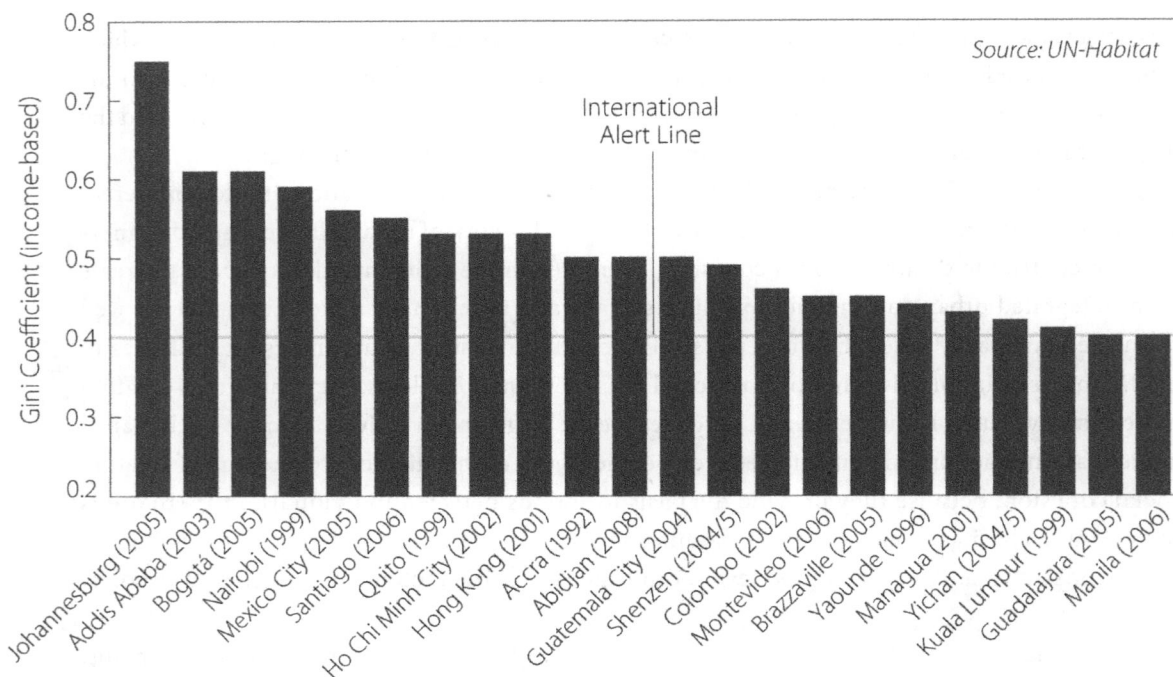

Figure 2.3.1 Most Unequal Cities by Income, Selected Cities in the Developing World, 1992–2008

In cities of developing and emerging countries, informal and illegal settlements accommodate up to 80 percent of the urban population, and the urban divide is often reflected in the spatial configuration of the city. But rising income inequality is also a challenge in developed countries. According to recent studies, the degree of segregation by income has risen in 11 of 13 major European cities, including Madrid and Vienna, as well as in 27 of the 30 largest major metropolitan areas in the United States, such as Houston and Los Angeles.[8]

Although the reasons for this trend are manifold, several key processes can be identified. In developed countries, the main factors driving segregation in cities are globalization, the withdrawal of government support, economic restructuring, and the lack of investment in social housing. The transition from a manufacturing to a service-based economy has led to a dramatic change in the job market. Fewer employment contracts are unlimited, many people work under precarious conditions and need more than one job to survive, and the service sector is not able to accommodate all the workers that lost their jobs in the context of de-industrialization.[9]

Moreover, cities and municipalities are cutting down on social expenditures and public services, while reducing or stopping investments in social and affordable housing, with the result being a rapidly decreasing low-income housing stock. This comes mostly at the expense of already disadvantaged population groups and exacerbates the separation of low- and high-income groups in urban areas.

Those residents who have the resources move to neighborhoods with better schools, while others who cannot afford the rising rents are displaced to the edge of the city. Consequently, global and local restructuring processes are closely intertwined and result in spatial patterns that reflect and accelerate inequality and exclusion in cities. (See Chapter 7.)[10]

The reasons for segregation in cities of developing and emerging countries relate mainly to the rise of the middle class, racial discrimination, provision of secure tenure, and economic liberalization, as well as to world-class city aspirations that often result in massive infrastructure and urban renewal projects with large-scale displacement of low-income or illegal groups. This prevailing trend of socioeconomic exclusion and spatial fragmentation has adverse consequences for the urban realm. As UN-Habitat explains: "[It] is impacting negatively on social cohesion and reduces the economic vibrancy and the overall prosperity of the city, including the quality of life of the citizens. Informal settlements and disconnected peripheries, dysfunctional public space and increasing insecurity are often the apparent results."[11]

Recognition of the negative impacts of exclusion and segregation calls for measures that "foster the development of a harmonious society in which all groups have a sense of belonging, participation, inclusion, recognition and legitimacy," according to researchers Gerard Boucher and Yunas Samad. Urban planners and designers can play an important role in this context, helping to support "inclusive cities" that value all people and their needs equally. The concept of inclusive cities often is approached through the lens of a particular marginalized group, such as the elderly, children, slum dwellers, migrants, the unemployed, or disabled people. Social cohesion is an important component of the inclusive city and is based on the notion of community building, cooperation, and social relations among persons of different socioeconomic and ethnic backgrounds. Urban planning and design measures at different scales can contribute to forming social ties and interaction—a prerequisite for social cohesion—and help to create a feeling of belonging in increasingly diverse and fragmented cities.[12]

National Urban Planning Programs and Frameworks

Socioeconomically deprived neighborhoods and city districts, which often are characterized by a high concentration of migrants and their descendants, cannot be understood in isolation. Their roots lie far beyond the local context. National and regional programs are needed to provide a framework for jump-starting local initiatives and to allocate financial means for these initiatives to work. As UN-Habitat has observed, "[r]ecent experiences have clearly shown that social integration, inclusion and cohesion can be promoted through interventions at different scales."[13]

National programs are only effective, however, if they are well-designed and are supported by institutional and governance structures. A review of four national planning programs implemented in Germany, Denmark, India, and South Africa demonstrates that numerous factors determine their success on the ground. (see Box 2.3.1.) These factors include: the selection process for deprived

Box 2.3.1   A Review of Four National Urban Planning Programs

Social City Program

Germany's Social City Program was established in 1999 with the objective of stabilizing and upgrading socially and economically deprived urban areas. It seeks to achieve social cohesion in often ethnically heterogeneous neighborhoods through an integrated approach that combines physical and social interventions in the target areas. As an important element of the federal urban development policy, the program was equipped with €150 million ($160 million) in 2015 (a significant increase from previous years) and had funded 659 actions in 390 cities as of the end of 2014.

National Urban Renewal Program

Based on the observation that poverty is increasingly urbanizing, South Africa's National Urban Renewal Program was established in 2001 as a 10-year initiative to promote socio-political, economic, and spatial integration of selected urban areas. The program focused primarily on exclusion areas (socially, economically, and racially) and supported eight urban districts in six cities, which were characterized by high levels of crime, poor connectivity to surrounding neighborhoods, high unemployment rates and inequality, and shortage of formal housing stock. Measures implemented under the program ranged from enhancing employment opportunities to enhancing access to the areas through better transportation services and improving education, local economies, and social capital.

Kvarterløft Program

Denmark's national urban regeneration program, Kvarterløft, ran from 1997 to 2007 and was later followed by the financially reduced Omradefornyelse. The area-based program was set up with the aim of addressing increasing social problems and the spatial concentration of immigrants and refugees. The program combined measures targeting both people and places and fostered coordinated and integrated approaches among different public sectors and by involving the local community.

Jawaharlal Nehru National Urban Renewal Mission

In contrast to the national programs mentioned above, India's Jawaharlal Nehru National Urban Renewal Mission did not apply an area-based approach but was launched in 2005 (and ran until 2015) with the goal of redeveloping entire cities and towns (65 in total) by making them more equitable, livable, and economically productive. With an investment of $20 billion, the program focused on upgrading infrastructure services and providing basic services to the urban poor. Its implementation faced numerous challenges, however, due to a lack of planners trained to realize integrated approaches, a shortfall in strengthening local governance, and a delay in financial flows from the national to the state governments.

Source: See endnote 14.

areas, the need for an integrated approach that combines physical and social measures, building local capacity, providing adequate financial resources, and conducting monitoring and evaluation.[14]

First, the selection process of targeted areas is critical to the success of interventions carried out at the local level. Yet the decision to declare a neighborhood "deprived" is often in different hands. In the case of South Africa and India, the national programs were centrally driven, which meant that the target areas were chosen top-down by either the national or state governments, without any consultation of local actors. Yet involving the local level and applying a bottom-up selection process—for example, by asking municipalities or communities to submit an expression of interest for participating in the program—are crucial for creating ownership. A proposal submitted by neighborhoods or cities simultaneously indicates awareness and their openness to change, thus increasing the chances of success in the long run. Both the German and Danish programs set up such an application process, which also helped them gain greater visibility and impact.

Second, the goal of social cohesion and interaction cannot be achieved solely through physical interventions, such as renovating residential buildings, improving lighting in public spaces, and reducing the number of housing units to combat vacancy. Rather, physical measures need to be combined with social measures aimed at improving living conditions in districts, such as creating new employment opportunities, providing better social and cultural facilities, and designing attractive public spaces that invite residents to stay and interact. The German program seeks to achieve exactly this: to upgrade the built environment while enhancing the situation of the local residents. Activities funded through the program range from modernization of buildings and the living environment; to supporting business start-ups, training, and education initiatives; to promoting language learning and fostering ethnic entrepreneurship and self-employment of immigrants and their descendants.[15]

Other programs, such as the Danish and South African ones, also highlight the need for combined social and physical efforts and place particular emphasis on the participation of local residents. On paper, such programs stress the need for combining physical and social interventions; however, their practical implementation remains controversial and widely criticized. Financially, infrastructure measures and physical upgrading have been the predominant focus of these programs, whereas social

Figure 2.3.2 Part of the Jawaharlal Nehru National Urban Renewal Mission (JNNURM) was funding for thousands of transit buses, including this one in Pune
Source: Rowan Vaz

initiatives and civic participation remain underrepresented. Denmark's Kvarterløft program, for example, claimed to focus on social initiatives and participation, yet more than 90 percent of the financial resources have been spent on physical improvements.[16]

Third, cities and municipalities often lack sufficient financial and personnel capacities and face weak coordination among different planning departments. The latter, in particular, poses a huge obstacle to the goal of simultaneously implementing social and physical measures, which can create fruitful synergies. For example, when designing a new public space, it would be advantageous to also consider how this could be coupled with providing space for local shops and a new community center, to create a vibrant place of interaction. However, coordination and communication is often insufficient, not only within the city administration, but also between different levels of administration. This can lead to different levels of administration having different understandings of the objectives to be achieved through the program, resulting in incoherence during implementation. The Indian and South African cases demonstrate that well-trained personnel, as well as structures and reforms at the local level, are needed to ensure that national objectives can be translated into local action.[17]

Fourth, because financial constraints often are major barriers to the implementation of concrete actions in cities, a national scheme that provides financial support for personnel and capacity building resources in cities can help in realizing concrete projects at the local level. Moreover, such a scheme can help finance the creation of the new institutions and agencies that may be necessary to manage and coordinate national programs on the ground. As part of the Social City program in Berlin, neighborhood management offices were gradually introduced in the target areas after 2005, with the overall goal of empowering local residents and involving them in decision-making processes and the development of their area. Neighborhood councils, consisting of and elected by local actors, decide how and for what projects the funds from the program can be used, and they also maintain the dialogue with the neighborhood management teams and the governmental administration. The neighborhood management offices facilitate networking and communication among existing nongovernmental organizations, businesses, and other social and cultural initiatives in the area to bundle and mobilize local resources.[18]

Fifth, monitoring and evaluation are key to tracking progress on the implementation of a program, identifying gaps, making adjustments where needed, and reviewing progress. However, systematic monitoring and evaluation often are not mandatory, and a lack of reliable data makes such initiatives difficult. Lessons learned from Denmark demonstrate the need to develop realistic indicators for measuring a program's progress and success, especially for aspects such as participation and empowerment, as well as establishing a continuous monitoring system.[19]

In sum, national programs can provide an important framework for local initiatives to work on the ground, yet respective mechanisms and structures must be in place to ensure greater impact. Although the four national programs discussed above were able to achieve positive change in the target areas, these were related mainly to upgrading the built environment, such as by renovating buildings and improving public plazas and other spaces. Social needs were often overlooked, and measures targeting the socioeconomic status of residents (for example, access to jobs, education, mobility, culture) were too limited. Many of the initiatives also had too short a time frame to create long-term change. It is crucial that national programs have a long-term scope and be based on continued political commitment. Interventions at the local level also need a scope that goes beyond the target areas in order to avoid stigmatizing them without ultimately remedying the situation.[20]

City-wide and Neighborhood Planning

National urban planning frameworks can—if designed properly—serve as a catalyst for local action to upgrade socioeconomically deprived urban areas. While national programs deliver pivotal framework conditions, their ultimate success rests on initiatives at the city and neighborhood levels. Key to designing sustainable and inclusive neighborhoods and cities is the ability to read and understand their language. Urban planners and designers need to carefully observe and analyze people's behavior in the urban realm and to design streets, public spaces, and entire neighborhoods accordingly. As author Jan Gehl puts it, to create "cities for people" or "people-friendly" cities, urban planning has to apply the human dimension that is focused on creating city spaces as *meeting places* for urban dwellers.[21]

A variety of planning and design measures allow for compact, well-connected, and integrated urban patterns that promote social cohesion in cities and provide spaces of encounter and social interaction. Among these are: land-use planning, the promotion of mixed-use areas with good access to public transport (via transit-oriented development), the rearrangement of street patterns, and public space design.

Land-use Planning for Balanced Urban Development

Land-use planning provides an important tool to guide and influence the development of cities. The consideration of not only economic aspects, but also environmental and social values, in land-use planning is necessary to allow for balanced and sustainable urban development. For example, community gardens fulfill important sociocultural functions and contribute greatly to social cohesion and food security. (See Box 2.3.2.) Yet in times of neoliberal city practices and enduring privatization of public land and properties in cities around the globe, such grassroots initiatives usually lack sufficient financial resources to continue.[22]

The recent wave of privatization has profound impacts on the urban realm and fails to acknowledge the increasing sociocultural complexity characterizing contemporary cities. It gradually diminishes the availability of spaces where new forms of social relations potentially could be formed. The selling of public assets, such as former school buildings, is often a shortsighted strategy that could cause unforeseen problems, as shown by the experience of Berlin. To consolidate the city's financial situation, the Berlin government established a property fund in 2001 to generate revenues through sales of city-owned land and properties in an auctioning process, without consideration of other aspects, such as the social value of initiatives. Some 400–500 public assets were sold annually, greatly reducing the number of city-owned properties. Yet when the unprecedented influx of refugees prompted a need for large-scale accommodations, the city government was forced to buy back buildings at much higher prices. Taking a more holistic and balanced approach in handling city-owned land and properties is key to preserving non-commodified spaces in cities and to retaining an adequate capacity to react in times of crisis. (See Chapter 16.)[23]

Box 2.3.2   Pro Huerta: Urban Agriculture and Food Security in a Changing World

As urban populations swell, many cities are struggling to ensure food security and adequate nutrition in the face of challenges such as climate change, economic and natural disasters, farmland degradation, and the immense barriers that the urban poor face to accessing fresh, nutritious food. Experiences with the Pro Huerta ("Pro Garden") program in Argentina and Haiti suggest that there are effective ways not only to improve nutrition, but also to shore up social resilience among vulnerable populations.

Buenos Aires and Rosario, Argentina

Argentina's National Institute of Agricultural Technology approved the Pro Huerta program in 1990 as a means to address the serious economic and food security challenges affecting the country, including a dramatic jump in food prices in Buenos Aires. Pro Huerta was formally adopted under the National Food Security Plan in 2003, and, in 2011, the government pledged more than $10 million to expand the program.

Pro Huerta helps Buenos Aires's poorest populations diversify their diets, access fresh food, lower their food budgets, and increase their incomes. The program is designed to boost self-sufficiency by providing the tools necessary to build food gardens, including seed kits, chickens and rabbits, and training in pest control, animal husbandry, and organic gardening methods. By late 2015, the program had helped set up more than 56,000 family gardens—supplementing the diets of some 350,000 people, or nearly 11.5 percent of the city's population—as well as more than 900 school gardens and 500 community gardens. A family garden can produce 200 kilograms of vegetables annually, enough for a five-person family.

Pro Huerta launched in Argentina's third largest city, Rosario, in February 2002. At that time, roughly 60 percent of the city's population was living below the poverty line, and food staples had quadrupled in cost, leading to theft and rioting. The Rosario government's Urban Agriculture Program and a local group, CEPAR, partnered to pilot the Pro Huerta model, offering tools and seeds to 20 gardening groups. By 2004, 800 community gardens were growing food for 40,000 residents.

The Pro Huerta program was successful in repurposing vacant land—comprising more than one-third of Rosario's land area—for gardens. The city has since updated its land-use laws to include urban farming and is building a green belt of parks and multi-scale gardens. Pro Huerta also has created venues for direct marketing to the public and has set up cooperatives that prepare and sell produce, soups, jams, and natural cosmetics. By 2004, 10,000 low-income households in Rosario were selling enough produce to lift themselves above the poverty line. An estimated two-thirds of participants were women.

In 2013, as the city's economy improved, participation dropped to some 1,800 residents, almost 14 percent of them full-time producers. The Pro Huerta program has been replicated in 88 percent of Argentina's municipalities, with more than 630,000 gardens and 130,000 farms providing food for over 3.5 million people nationally. A network of 20,000 promoters manages the program, participating in agro-ecological fairs and working with thousands of institutions and organizations across Argentina.

Haiti

The Pro Huerta program also has spread to Brazil, Colombia, Guatemala, Venezuela, and Haiti, a country that suffers from widespread poverty, inadequate nutrition, and high dependence on food imports. Haiti launched Pro Huerta in 2005 with support from the Argentine Fund for Horizontal Cooperation, adapting the program to the local context and using local leadership to manage it. Argentinian experts trained a team of Haitian agricultural engineers, who then taught a network of volunteer promoters—mostly women—how to provide trainings within their communities. Between 2005 and 2008, these efforts helped establish 16,086 family gardens, 2,700 school gardens, and 1,900 community gardens.

In addition to producing food, Pro Huerta has resulted in the creation of resilient social networks that have helped communities respond to disruptions. In 2008, after Hurricanes Gustav and Ike destroyed thousands of gardens across Haiti, the program bounced back thanks to robust community cohesion. By the end of 2009, 1,843 promoters, 11,465 gardens, and more than 80,000 participants were active in the program. Following the 2010 Haiti earthquake, Pro Huerta was instrumental in fighting the cholera outbreak, providing more-nutritious diets for susceptible populations, offering expertise on food handling, and building special water storage facilities and sand filters to avoid disease transmission.

According to a Pro Huerta survey, 93 percent of program participants in Haiti improved their food situation, 86 percent of households were able to access a greater variety and quantity of food, and household spending was halved to just 33 percent of monthly income. In 2014, the Union of South American Nations pledged $3 million to extend Pro Huerta to 2016, with the goal of nearly doubling participation to 220,000. Haiti hopes to extend the program to 1 million participants by 2019.

Lessons Learned

In Argentina, the Pro Huerta program was predominantly a response to short-term economic disturbance; however, the model also performs strongly in countries, such as Haiti, that face continuous threats to food security. In a world where food supply and access are increasingly affected by variations in climate, environmental conditions, equity, and natural and economic disasters, urban agriculture programs such as Pro Huerta can be used to empower underserved communities, providing them with the tools to build a healthier life and to help them cope with future turmoil and change.

Kristina Solheim, Program Manager, goNewHavengo

Source: See endnote 22.

Integrating Land-use and Transport Planning to Foster Social Cohesion: The Example of Transit-oriented Development (TOD)

At the neighborhood level, integrated land-use and transport planning—coupled with the creation of high-density mixed-use areas—facilitates demographic, socioeconomic, and cultural diversity. In particular, transit-oriented development (TOD) has become a popular planning approach to create inclusive, connected communities through spatial planning. Regulatory and incentive mechanisms, such as local planning schemes, educational campaigns, and incentives for developers and communities, are crucial for successful implementation. TOD is based on the principle of designing high-quality mixed-use areas around transit stations to enhance access to public transport and pedestrian- and cyclist-friendly environments while reducing dependence on private cars. Areas that prioritize walking and cycling typically are characterized by higher levels of social interaction and help residents who are unable to afford a car to overcome transport poverty. (See Chapter 11.)[24]

The design of mixed-use areas follows the idea of creating "urban villages" where residents are provided with housing, transportation, community and recreational facilities and services, public spaces, and retail within a short distance. To facilitate community diversity and social cohesion, these services and facilities should cater to the needs of different social groups with varying interests and demands. TOD therefore should be designed and managed in a way that allows for diversity in housing (for example, in design, form, tenure, and affordability), land-use, employment, and retail, and that provides multiple public and open spaces as focal points for the community. Safeguarding community diversity over the long run requires long-term investments in social housing and community infrastructure. Further, developing a TOD precinct requires a continuous participatory planning process that targets a diversity of groups to build ownership and a shared sense of identity.[25]

Well-designed TOD offers numerous environmental benefits. The continued rise in transport volumes not only leads to increasing traffic congestion, but also contributes to environmental and health challenges such as rising air pollution and greenhouse gas emissions. Designing neighborhoods based on the principle of walkability and cycling as well as good access to public transport is urgent. TOD holds tremendous potential in countries like China. (See Chapter 7.) The country's third largest city, Guangzhou, has invested massively in a highly efficient bus rapid transit (BRT) system and is building new promenades and bicycle lanes to encourage walking and cycling.

Guangzhou's BRT system, which is the first worldwide that is fully integrated with a metro system, carries more than 800,000 passengers daily and has significantly reduced traffic jams and vehicle kilometers traveled. Thanks to multiple sub-stops and passing lanes at each station, average bus speeds increased from about 15 kilometers per hour to about 22 kilometers per hour—an attractive and speedy alternative to individual motorized transport. The network of small, walking-oriented streets surrounding the city's Shipaiqiao station is being complemented by new high-density commercial

Figure 2.3.3 Dedicated BRT lanes on Tianhe Road, Guangzhou
Source: David290

and residential developments, helping to revitalize the entire area. Shipaiqiao station and its surrounding area are now easily accessible by public transport and have become a prime location for shopping, working, living, and strolling.[26]

Upgrading Street Networks to Reintegrate Neighborhoods

Streets can have a great impact on the vitality and integration of a given area. They are not only a means of transportation, but also a fundamental shared public space that facilitates numerous social, cultural, and economic activities and allows people to interact. Well-designed street patterns that facilitate connectivity and mobility can counteract socio-spatial segregation and help to re-integrate areas into city structures.

UN-Habitat promotes a street-led approach to the citywide transformation and regeneration of slum areas in many developing and emerging countries. The absence of streets and open spaces segregates and disconnects slums from the rest of the city. Done right, the upgrading of street networks in slums can bring advantages including security of land tenure, future consolidation of settlements, optimization of land use, poverty reduction, and increased social interactions among residents. However, upgrading of street networks also requires political will and needs to be based on a strong participatory planning process. The latter is a necessary precondition not only to create ownership, but also to conduct a reliable inventory of the physical configuration and socio-spatial structure of a settlement. Participatory planning also helps to inform the design of area-based plans and street patterns that capture the "multiple functions of streets based on nuances of everyday practices of street life and people's aspirations."[27]

The Integrated Program for the Improvement of Squatter Areas (PRIMED) in Medellín, Colombia, provides a good example of the benefits of well-designed street networks. Based on strong political will and the desire to counteract spatial exclusion and promote social development in deprived areas, the program facilitated the application of an innovative public transport system based on cable cars connecting the target areas with the rest of the city. (See Chapter 4.) The first cable car line was implemented in the poor and densely populated northeastern district, which was characterized by minimal road infrastructure and thus a lack of accessibility. The program was not limited to the implementation of the transport system, however, but also combined urban upgrading measures, including interventions in public spaces, social housing, and other social infrastructure, which were realized in a participatory manner.[28]

Impact studies reveal that these combined interventions helped to upgrade Medellín's densely populated and low-income neighborhoods and integrate them into the city's fabric. They also boosted the quality of life of the urban poor by enhancing accessibility for local residents and outsiders alike, improving air quality, counteracting stigmatization of these areas, and providing local residents with a sense of social and political inclusion. Levels of violence and crime in the neighborhoods surrounding the cable car lines dropped significantly, which helped to revitalize public life. In addition to social and mobility aspects, the PRIMED program considered environmental outcomes. The Metro Company, which evaluates the environmental performance of the cable car system and monitors the reduction in greenhouse gas emissions, concludes that the hydroelectric aerial cable cars could help to reduce up to 121,029 tons of carbon dioxide between 2010 and 2016, compared to the fossil fuel-operating vehicles that the system replaces.[29]

Although the upgrading of street networks can support development, foster integration, and bring environmental benefits, it often requires demolition and relocation to make space for the construction of new streets or an aerial cable car public transport system. This tradeoff was evident in the case of Medellín and had to be negotiated within the community. Overall, however, improved street networks have great potential to integrate entire neighborhoods into city systems and to improve the quality of life of local residents.

Public Space Design

Public spaces allow people to meet and interact on ostensibly neutral ground. They provide a democratic space for different social groups to participate in civic activities. Especially in developing and emerging countries, where urban inhabitants often live in densely populated housing areas with few economic resources, public spaces form a fundamental part of community life. Urban parks, in

Figure 2.3.4 A sunny July day on the waterfront promenade at Aker Brygge in Oslo
Source: Jean-Pierre Dalbéra

particular, serve as a vital public space where everyday experiences are shared and negotiated among different social and ethnic groups, and where numerous opportunities for intercultural interaction exist.

Extensive city improvements, such as upgrading street networks, are costly and time-consuming. However, small interventions in public spaces—such as improvements to bench seating, providing movable chairs, closing streets to car use, and laying new pavement to encourage pedestrian traffic—can make a huge difference and help reinforce daily life in a fast and cost-efficient manner. For example, the location of street furniture has a compelling effect on how public space is used and accepted and how long people tend to stay and interact with strangers.[30]

The post-industrial waterfront promenade at Aker Brygge in Oslo is a good example of how urban design can influence social interaction. As part of a broader neighborhood renovation project in the 1990s, old benches on the promenade were replaced with Parisian-style double park benches, and the overall seating capacity was increased. Consequently, the number of people sitting in the area more than doubled, and social interactions among strangers multiplied. Some two decades later, the same architects were tasked with adapting and renewing the area, again with an emphasis on encouraging social interaction and diversity. They developed a "site-specific concept for street furniture and 'staying,'" which aided in creating numerous opportunities to sit, lie, eat, read, or chat with acquaintances or strangers. The pedestrian and bicycle path was reorganized to create wider, more generous public spaces, and sun loungers and comfortable benches were installed, inviting people to sunbathe and lie down. Although the provision of sufficient seating opportunities within cities and neighborhoods is crucial, other factors—such as views and orientation toward street activities, as well as movability of seating options—determine the vitality of a place.[31]

Conclusion

Socioeconomic polarization and spatial segregation have become prevailing trends in cities worldwide, with adverse impacts on quality of life and social cohesion. As cities become increasingly diverse, these trends often have an ethnic component as well. Many socioeconomically deprived areas are characterized by a high concentration of migrants, making their multi-faceted integration into city life more challenging. Consequently, finding solutions to counteract disparities and inequalities while strengthening relations and interactions among socially and ethnically diverse groups has become an urgent matter.

Although urban planners and designers cannot solve the roots of exclusion and inequality per se, they can aid in increasing the accessibility and integration of deprived areas and provide spaces that increase the chances of interaction and the forming of social relations among people from differing ethnic backgrounds. National urban planning programs offer a useful framework for local initiatives to kick off and work on the ground. Applying an integrated approach that effectively combines social and physical measures, coupled with a bottom-up selection process, capacity building, the establishment of governance structures, the provision of financial resources, and monitoring and evaluation is key for the success of national programs.

At the city and neighborhood levels, numerous approaches and measures have been tested globally to overcome socio-spatial segregation and exclusion. In particular, the creation of mixed-use and socially mixed areas—coupled with good access to public transport, housing diversity, and sufficient provision of vibrant public spaces that facilitate inter-ethnic encounters—are promising ways to enhance social cohesion. Approaches and planning principles, such as socioeconomically balanced land-use planning, transit-oriented development, and upgrading street patterns, have been successful in building well-connected, compact, and integrated urban patterns that allow for sustainable urban development. Well-designed public spaces also can serve as a key locus where new forms of sociability can emerge. Urban planners and designers have the tools and instruments at hand to contribute greatly to social cohesion in cities, yet political will and the participation of a broad array of stakeholders, including local residents, is a fundamental precondition to the success of any measure.[32]

Endnotes

1. UN-Habitat, *State of the World's Cities 2008/2009: Harmonious Cities* (Nairobi: 2009); United Nations Department of Economic and Social Affairs (DESA), Population Division, *World Urbanization Prospects: The 2014 Revision, Highlights* (New York: 2014); International Organization for Migration (IOM), *World Migration Report 2015. Migrants and Cities: New Partnerships to Manage Mobility* (Geneva: 2015).

2. Mary J. Hickman and Nicola Mai, "Migration and Social Cohesion. Appraising the Resilience of Place in London," *Population, Space and Place* 21, no. 5 (2015): 431.

3. UN-Habitat, *State of the World's Cities 2012/2013: Prosperity of Cities* (London: Routledge, 2013), 150.

4. U.K. Department for Communities and Local Government, *English Housing Survey 2010 to 2011: Headline Report* (London: 2012); Patrick Butler, "'Inadequate, Unaffordable, Insecure': UK Housing's Decline and Fall," *The Guardian* (U.K.), September 11, 2013.

5. Karin Peters, Birgit Elands, and Arjen Buijs, "Social Interactions in Urban Parks. Stimulating Social Cohesion?" *Urban Forestry & Urban Greening* 9, no. 2 (2010): 93–100.

6. UN-Habitat, *Urban Planning and Design for Social Cohesion. Concept Note World Urban Forum* (Medellín, Colombia: April 2014), 2; IOM, *World Migration Report 2015*, 4.

7. Gerard Boucher and Yunas Samad, "Introduction. Social Cohesion and Social Change in Europe," *Pattern of Prejudice* 47, no. 3 (2013): 197; UN DESA, United Nations Development Programme (UNDP), and Office of the United Nations High Commissioner for Human Rights (OHCHR), *Habitat III Issue Papers—1 Inclusive Cities* (New York: 2015). Figure 2.3.1 from UN-Habitat, *State of the World's Cities 2010/2011: Cities for All—Bridging the Urban Divide* (New York: 2011), 73.

8. UN DESA, UNDP, and OHCHR, *Habitat III Issue Papers*; Tiit Tammaru et al., eds., *Socio-Economic Segregation in European Capital Cities. East Meets West* (London: Routledge, 2015); Richard Fry and Paul Taylor, *The Rise of Residential Segregation by Income* (Washington, DC: Pew Research Center, August 1, 2012).

9. Tammaru et al., eds., *Socio-Economic Segregation in European Capital Cities.*

10. Hartmut Häussermann, "Wohnen und Quartier: Ursachen sozialräumlicher Segregation," in Ernst-Ulrich Huster, Jürgen Boeckh, and Hildegard Mogge-Grothjahn, *Handbuch Armut und soziale Ausgrenzung* (Wiesbaden: VS, 2008), 335–49.

11. Jane Parry, *Issue Paper on Secure Tenure for Urban Slums. From Slums to Sustainable Communities: The Trans-formative Power of Secure Tenure* (Atlanta and Brussels: Habitat for Humanity and Cities Alliance, 2015); UN-Habitat, *Urban Planning and Design for Social Cohesion*, 2.

12. Boucher and Samad, "Introduction. Social Cohesion and Social Change in Europe"; Peters, Elands, and Buijs, "Social Interactions in Urban Parks"; Talja Blokland, Carlotta Giustozzi, and Franziska Schreiber,

"The Social Dimensions of Urban Transformation: Contemporary Diversity in Global North Cities and the Challenges for Urban Cohesion," in Harald A. Mieg and Klaus Töpfer, *Institutional and Social Innovation for Sustainable Urban Development* (Oxon and New York: Routledge, 2013), 125.

13. UN-Habitat, *Urban Planning and Design for Social Cohesion*, 1.

14. Box 2.3.1 from the following sources: German Federal Ministry for the Environment, Nature Conservation, Building and Nuclear Safety, *Social City Program* (Berlin: 2015); Alexandra Galeshewe et al., *National Urban Renewal Programme. Implementation Framework* (Pretoria: Department of Provincial and Local Government, Republic of South Africa, undated); Michael E. Leary and John McCarthy, *The Routledge Companion to Urban Regeneration* (London and New York: Routledge, 2013), 402; Hans Skifter Andersen and Louise Kielgast, *Area-based Initiatives in Denmark—"Kvarterløft": Addressing Increasing Social Problems and Concentration of Immigrants and Refugees in Seven Neighborhoods* (Copenhagen: Danish Building Research Institute, June 2003).

15. German Institute of Urban Affairs, *Status Report. The Programme "Social City" (Soziale Stadt)—Summary* (Berlin: Federal Ministry of Transport, Building and Urban Affairs, 2008).

16. Galeshewe et al., *National Urban Renewal Programme*; Thomas Franke and Wolf-Christian Strauss, *Management gebietsbezogener integrativer Stadtteilentwicklung. Ansätze in Kopenhagen und Wien im Vergleich zur Programmumsetzung "Soziale Stadt" in deutschen Städten* (Berlin: German Institute of Urban Affairs, 2005).

17. Franke and Strauss, *Management gebietsbezogener integrativer Stadtteilentwicklung*; Ivan Turok, *The Evolution of National Urban Policies: A Global Overview* (Nairobi: UN-Habitat and Cities Alliance, 2014).

18. Senatsverwaltung für Stadtentwicklung, "The Neighborhood Council Within the Neighborhood Management Process," handout at the 3rd Congress of Berlin's Neighborhood Councils (Berlin: March 20, 2010).

19. Franke and Strauss, *Management gebietsbezogener integrativer Stadtteilentwicklung*.

20. Ellen Højgaard Jensen and Asger Munk, *Kvaterløft. 10 Years of Urban Regeneration. Ministry of Refugees, Immigration and Integration Affairs* (Copenhagen: 2007).

21. Jan Gehl, *Cities for People* (Washington, DC: Island Press, 2010).

22. Chris Firth, Damian Maye, and David Pearson, "Developing 'Community' in Community Gardens," *Local Environment: The International Journal of Justice and Sustainability* 16, no. 6 (2011): 555–68. Box 18–2 based on the following sources: International Network for Economic, Social & Cultural Rights, "Report and Recommendation on Request for Inspection, Re: Argentina—Special Structural Adjustment Loan 4405-AR (Pro-Huerta Case)," 2012, https://www.escr-net.org/node/364789; Ana Bell, "Community Gardens Boost Self-sufficiency in Argentina," Panos London, August 31, 2012; Ministry of Foreign Affairs and Worship of Argentina, "Desarrollo sustentable: Haiti—autoproduccíon de alimentos frescos Pro Huerta," http://cooperacionarg.gob.ar/en/haiti/autoproduccion-de-alimentos-frescos-pro-huerta;

Instituto Nacional de Tecnología Agropecuaria (INTA), "Pro Huerta," http://prohuerta.inta.gov.ar/; Walter Alberto Pengue, "Aún nos quedan las manos y la tierra," *El Diplo* 38 (August 2002); Municipality of Rosario, "Indicadores Demograficos," November 23, 2015, www.rosario.gov.ar/sitio/caracteristicas/indicadores. jsp; United Nations Food and Agriculture Organization, "Rosario," in *Growing Greener Cities in Latin America and the Caribbean* (Rome: 2013); Ferne Edwards, "Sustainable City & Model—Urban Agriculture in Argentina," Sustainable Cities Network, July 13, 2007, www.sustainablecitiesnet.com/models/model-urban-agriculture-in-rosario-argentina/; Ministry of Social Development of Argentina, "Pro Huerta," 2013, www.desarrollosocial.gob.ar/wp-content/uploads/2015/08/1.-M--s-sobre-PRO-HUERTA. pdf; Canadian International Development Agency and Inter-American Institute for Cooperation on Agriculture (IICA), "Argentina, Canada and Haiti Join Efforts to Improve Food Security. Project for Self-sufficiency in the Production of Fresh Foods in Haiti Is Expanded," press release (Haiti: June 2008); IICA, "Program for Fresh Food Self-sufficiency in Haiti: Pro-Huerta 2005-2008," *Comuniica*, January–April 2008; Pan-American Health Organization and Ministerio de Relaciones Exteriores, Comercio Internacional y Culto de la República Argentina, *South-South Cooperation: Triangular Cooperation Experience Between the Government of the Argentine Republic and the Pan-American Health Organization/World Health Organization* (Buenos Aires: October 2009); "Lessons Learned in Argentina Helping Haiti Cope with Cholera," *New Agriculturalist*, December 2010; "Haiti Agriculture: True Success of Pro Huerta Program in Haiti," *Haiti Libre*, March 23, 2015; "Haiti—Agriculture: The Argentinean Program Pro Huerta Extended Until 2016," *Haiti Libre*, January 17, 2014.

23. Jacqueline Groth and Eric Corjin, "Reclaiming Urbanity: Intermediate Spaces, Informal Actors and Urban Agenda Setting," *Urban Studies* 42, no. 3 (2005): 503–26; David Harvey, *Rebel Cities: From the Right to the City to the Urban Revolution* (London and New York: Verso, 2012); Franziska Schreiber, "Viele viele Frei(t)räume: The Prinzessinnengarten and Contemporary Land Use Conflicts in Berlin," anstiftung. de/downloads/send/15-forschungsarbeiten-urbane-gaerten/173-the-prinzessinnengarten-and-contemporary-land-use-conflicts-in-berlin.

24. The Queensland Government, *Transit Oriented Development: Guide to Community Diversity* (Brisbane: Queensland Department of Infrastructure and Planning, 2010); Gehl, *Cities for People*; Xuemei Zhu et al., "A Retrospective Study on Changes in Residents' Physical Activities, Social Interactions, and Neighborhood Cohesion After Moving to a Walkable Community," *Preventive Medicine* 69, no. 1 (2014): 93–97.

25. Gehl, *Cities for People*, 7.

26. Institute for Transportation and Development Policy—China, *Best Practices in Urban Development in the Pearl River Delta* (Guangzhou: December 2012), 81–88.

27. UN-Habitat, *Streets as Tools for Urban Transformation in Slums. A Street-led Approach to Citywide Slum Upgrading* (Nairobi: 2012), 15.

28. Julio D. Dávila and Diana Daste, "Aerial Cable-Cars in Medellín, Colombia: Social Inclusion and Reduced Emissions," in Mark Swilling et al., *City-Level Decoupling: Urban Resource Flows and the Governance of*

Infrastructure Transitions. Case Studies from Selected Cities. A Report of the Working Group on Cities of the International Resource Panel (Paris: United Nations Environment Programme, 2013), 47–48.

29. Ibid.

30. Gehl, *Cities for People*, xii.

31. Link Arkitektur, "Stranden—Aker Brygge," http://linkarkitektur.com/en/Projects/Stranden-Aker-Brygge.

32. Justus Uitermark, "'Social Mixing' and the Management of Disadvantaged Neighbourhoods: The Dutch Policy of Urban Restructuring Revisited, *Urban Studies* 40, no. 3 (2003): 531–49.

DISCUSSION QUESTIONS

1. What are the most significant environmental challenges in cities these days? Are the problems the same in the developed world and developing world?

2. What is the notion of social cohesion? How has the concept experimented with? Explain the approach to enhancing social cohesion in national urban planning programs.

3. Do you believe the design of mixed-use areas could facilitate community diversity and social cohesion? What are the practical mechanisms to realize mixed-use areas to promote community diversity and social cohesion?

4. Transit-oriented development (TOD) has become a popular planning approach to create inclusive, connected communities through spatial planning. It is considered beneficial since it can not only enhance social cohesion but also offer environmental benefits. What are the environmental benefits? Do you think the TOD could also generate unintended adversarial environmental impacts?

Renewable Energy Delivery and Expansion with Public and Private Partnerships for the Global South

By Kyoo-Won Oh and Younsung Kim

Introduction

Energy is necessary for individuals' daily lives, a foundation of the shared activities in a society, and an important source of economic growth in a country in this increasingly modernized world. The provision of stable energy in a country is thus one of the top priorities for policymakers as an independent energy policy and an important cross-cutting policy that affects the entire economy and its different industrial sectors (Chaurey et al., 2012).

In pursuing energy policy, the energy supply issue has been confounded by climate change concerns these days. With the increasing scientific evidence for climate change (IPCC, 2018), the international policy community has agreed to limit global mean temperature increase below 1.5°C at the Paris climate conference (UNFCCC, 2016). Before and during the conference, both developing and developed countries submitted comprehensive national climate action plans, and the plans included renewable-based energy supply as a climate mitigate action. Evidently, renewable-based energy system transition and expansion can be an ideal way to address a twofold energy provision challenge, meeting increased energy access needs, particularly in the Global South, and transitioning to a clean energy system to address climate change.

Among various policy support mechanisms, well-designed public–private partnerships (PPPs) could be one of the potentially viable options for providing renewable-based energy services in developing countries (Kruckenberg, 2015; Martins et al., 2011). In the face of insufficient government investments due to the limited public budgets, PPPs would overcome the budgetary constraints of capital-intensive renewable energy projects demanding high up-front costs and inducing high efficiencies in the operation and maintenance of renewable energy. PPPs would also enable public entities to gain access to private sectors' cutting-edge technical expertise, innovative financing solutions, and efficient delivery of public goods in a timely manner, essential in the stages of project development, operations, and maintenance. However, most energy projects in developing countries had traditionally relied on single-handedly public efforts. The government-initiated projects often led to project delays or failures when they are less capable of ensuring affordable service fees for the poor, drawing community engagement and establishing well-functioning institutional management systems (Komendantova et al., 2012; Martins et al., 2011; UNDP 2011). In this light, PPPs can be a handy tool for public managers or policymakers in developing countries to achieve energy access goals with renewable energy.

The current literature on PPP has rarely discussed the significance of PPPs for increased energy access in the context of developing countries. Some prior studies examined the application of PPP arrangements in the energy sector, but they are limited in that their focus is mainly on developed nations (Dinica, 2008; Martins et al., 2011) or large-scale projects in urban areas (Komendantova et al., 2012). In this article, we first discuss the overall benefits and drawbacks of renewable energy development. Furthermore, details of PPP arrangements and risk allocations for renewable energy development are also examined. We then explore a case of a 5 megawatts (MW) mini-size rooftop solar project in Gujarat, India, and discuss what makes this case proven to be successful. In doing so, our study indicates that if a PPP is well-designed, it can be an efficient, effectively affordable mode of small-scale electrification for sustainable energy supply in developing and emerging economies.

Benefits and Drawbacks of Renewable Energy Development

Renewable energy is defined as an energy source "derived from natural processes that are replenished constantly" (IEA, 2002). Its various forms include electricity and heat generated from solar, wind, ocean, hydropower, biomass, geothermal resources, and biofuels, and hydrogen drawn from renewable resources. Given the non-depletion of resources and less harmful environmental impacts, renewable energy has received broad support internationally compared to fossil fuels or nuclear energy. However, promoting renewable energy development in a country would eventually depend on various factors. For instance, a country's concerns over fossil fuel reserves, imported fossil fuels price fluctuations, cost-ineffectiveness in existing power infrastructure, and high climate risks would create a favorable policy environment, resulting in a strategic direction toward renewable energy production (Dinica, 2008). Political stability and policy support would also bring more renewable projects in developing and emerging countries (Komendantova et al., 2012).

Renewable energy development provides various benefits. First, it strengthens energy security and independence through a diversified energy mix with locally produce energy, therefore enabling a stable power supply to various economic activities in a country. Second, renewable energy allows a country to contribute to slowing down the threat of climate change by lowering greenhouse gas emissions during electricity generation and avoiding some of the related economic losses caused by climate change. Third, new business and job opportunities in various sectors would be created during the deployment of new renewable-based technologies, which would positively shape national competitiveness and economic growth.

However, there are several drawbacks to renewable energy development. First, while renewable technologies have made great leaps, their power generation efficiencies remain still lower than those of conventional fossil fuel–fired power generation. As such, in order to promote renewable project development, governments have adopted diverse forms of financial support mechanisms such as fiscal incentives, direct and indirect subsidies, tax exemptions, and soft loans. In particular, feed-in tariffs (FITs) have been well-favored by many developed countries, including Germany. The FIT scheme provides legal guarantees for long-term purchase contracts with utility companies at a fixed attractive price (Dinica, 2008), leading private investors to a high level of assurance on the financial projection of renewable energy projects and thus facilitating renewable energy market generation (Komendantova et al., 2012).

Second, renewable energy projects pose a capital-intensive cost structure where high up-front costs are required at the stages of project development and construction (Sovacool, 2011). High capital costs incur as renewable energy projects include a supply-chain bottleneck where there are still fewer companies with required renewable technologies than the existing demands. Moreover, additional costs are required to connect to the electricity grid, when in most cases, renewable energy plants are located remotely and distant from end-users. Renewable energy intermittency, for example, daytime for solar energy, the wind blowing for wind energy, and rainy season for hydro energy, is also translated into high costs (Komor, 2004). Therefore, it is widely perceived that renewable energy is not cost-competitive as compared to fossil fuels. However, the true costs of fossil fuel–fired power generation should be higher than the currently projected ones when environmental and social costs concerning negative externalities (e.g., industrial pollution and natural resource depletion) are captured. Fossil fuel subsidies also distort the true costs of fossil fuel–based power generation, lowering actual renewable-based electricity generation costs (Bridle & Kitson, 2014).

Renewable Energy Projects in Developing Countries

In many parts of the world, around 1.4 billion people still do not have access to electricity. More than 2.6 billion people in developing countries rely on traditional biomass for cooking and heating (World Bank, 2014). This lack of access to modern energy services continues to impede sustainable development. It also reflects an imminent need for installing modern energy service systems in developing countries, especially in South Asia and Sub-Saharan Africa (REN 21, 2016). If it is developed in a way to ensure financial affordability and technical sustainability, renewable energy has a huge potential for meeting these countries' energy needs, especially when combined with the use of mini-grids in un-electrified peri-urban and rural areas. Thus, facilitating energy access through renewables has become an influential policy agenda in developing countries (Chaurey et al., 2012, REN 21, 2016; UNDP, 2009).

Against this backdrop, the renewable energy market has grown exponentially in the last decade and continues to build its global electricity market share. Renewables such as modern biomass, solar, geothermal, wind, tidal, and wave, but excluding large hydro projects, accounted for 7.1% of the total world electricity generation in 2013 (REN 21, 2016), as compared to 2% in 2004 (REN 21, 2005). With the total investment of $241.6 billion from public and private entities, renewable energy accounts for 55% of the new generating capacity installed worldwide in 2013 (UNEP, 2017). Renewable energy investment in developing countries has also shown a drastic increase from $8 billion in 2004 to $116.6 billion in 2016, mostly in Latin America and the Caribbean and some developing Asia.

From the analysis of around 1,700 renewable energy projects from 1990 to 2012 on Private Participation in Renewable Energy (PPRE) Database of World Bank (2014), we identified several trends as follows. First, the wind has been the most active technology deployed since 1990. In 2012, there were 25,954 MW of wind energy projects with private participation that reached financial closure in developing countries, and total project costs were about $46 billion. Many of the wind energy projects have been developed in Latin America and the Caribbean. At the same time, East Pacific Asia showed the least total electrification capacity in both financed and pipeline wind power projects. Solar technology ranks second, and its rapid growth in complete pipeline projects was recently made for the past five years. In 2012 alone, 1,798 MW of PPP solar projects reached financial closure in developing countries, and their total project costs were estimated at $7.5 billion. The Sub-Saharan region has been very active in attracting foreign private investments in renewable energy projects, and South Africa has been the most active in the region. Chile has also shown the highest commitment to creating favorable energy, as solar projects with nearly 1,739 MW of total solar capacity were installed. In doing so, Chile almost doubled South African's installments' full capacity in 2012.

However, geothermal and other renewables (wave and tidal) have not been actively tapped, resulting in a small number of pipeline and financed projects. Geothermal projects were proposed in a few countries such as Indonesia, Bangladesh, and Malaysia, with their total capacity being 840 MW since 1990. In 2011, there was only one wave energy project proposed off the Kenyan coast with a capacity of 100 MW concerning other technologies. The development of biomass energy has also been slow, and its contribution is still a marginal component of the total renewable energy supply. Only 3% of the total financed renewable projects were based on biomass energy, and 582 MW of projects with private participation reached financial closure in developing countries. Most biomass energy projects proposed and implemented were limited in five countries, China, Brazil, Romania, India, and Uganda.

Regional disparity is one of the critical features in renewable energy development of the developing world. Latin America and the Caribbean, and East Asia and Pacific are the most active regions in renewable energy supply, while the African region with the lowest electrification rate still suffers from the fewer pipeline and financially closed renewable energy projects. This uneven regional distribution would be in part attributed to the availability of subsidy-based incentives offered to project developers, along with relatively low political risks to private investors and a higher level of bankability in large middle-income countries, including China, India, and Brazil.

Our analysis of PPRE also indicates that most PPP projects in developing countries were promoted by federal governments' initiatives, reflecting the centralized governance system for energy development and planning in developing countries. Noticeably, around 80% of renewable projects were contracted by federal governments. In comparison, state and local governments made only 5% of PPP arrangements, mainly in India, Thailand, China, Malaysia, and Mexico. In contrast, PPP-based renewable energy development in developed countries tends to constitute a meta-governance system by involving federal, state, local governments, and sometimes independent legal bodies incorporated in a country (Koch & Buser, 2006).

Public–Private Partnership Arrangement in Renewable Energy Development

PPP is a procurement method where private and public sectors jointly undertake a project to provide public goods or facilities, which otherwise would have been undertaken only by the public sector (Martins et al., 2011). The public partner can take different forms, such as the central and local government, or even independent legal bodies incorporated in the state or regional and local administration (Koch & Buser, 2006). In traditional government procurement, the private sector delivers only certain limited tasks required for a larger project, while the government's responsibility is to monitor and consolidate the various works undertaken by different private sector companies for larger system delivery. Under a PPP structure, the private sector resumes a broader range of responsibilities depending on how project risks are allocated between the public and the private sector (Komendantova et al., 2012).

Roles of the Private Sector in Renewable Energy PPP

In many developing countries, power is usually provided by a central government through state-owned-enterprises (SOE) for power generation, transmission, and distribution. However, limited public resources would lead the government not to be able to meet the growing demands of electricity alone, therefore a strong need to mobilize monetary resources from the private sector. This is especially so for the generation of renewable energy, which requires high capital costs and a high level of technological know-how during project development, construction, and operations. Furthermore, the mismatch between the up-front high capital cost requirements of renewable energy and revenue generation over a long-term period for the cost recovery needs to be bridged by innovative financing solutions with complex legal documents and proper risk allocations. However, these technical, operational, and financial skills required for renewable energy development are not usually of the government but the private sector.

Typical PPP Structure of Renewable Energy Project

Renewable energy PPP project also follows a typical PPP structure, where a government grants a concession to private sector investors for a certain period. In turn, the investor, with required responsibilities, establishes a Special Purpose Company (SPC) and makes necessary arrangements to raise required funds to undertake the project through equity investments and debt instruments. The investor recovers its investment with a reasonable return and repays the debts from a series of revenues generated from the project over the long-term concession period. In the case of renewable energy PPP projects, like any other power PPP projects, investors' economic and financial risks are reduced or eliminated to a certain extent through a long-term Power Purchase Agreement (PPA). The PPA is usually signed between the SPC, which generates and sells electricity, and a utility company that purchases electricity from the SPC. The commercial terms for the sales of electricity, for example, the price and volume, conditions of payments, responsibilities of each party, contract period, and termination events, are defined in the PPA. Due to the low credibility of the utility SOE in developing countries, regular and termination payments and other key responsibilities in the PPA are often guaranteed by the government. The core of PPP arrangement lies in identifying and allocating the risks embedded in a project among various participating parties, which are considered to be best in handling the risks. A typical structure of renewable energy PPP project is illustrated in Figure 2.4.1.

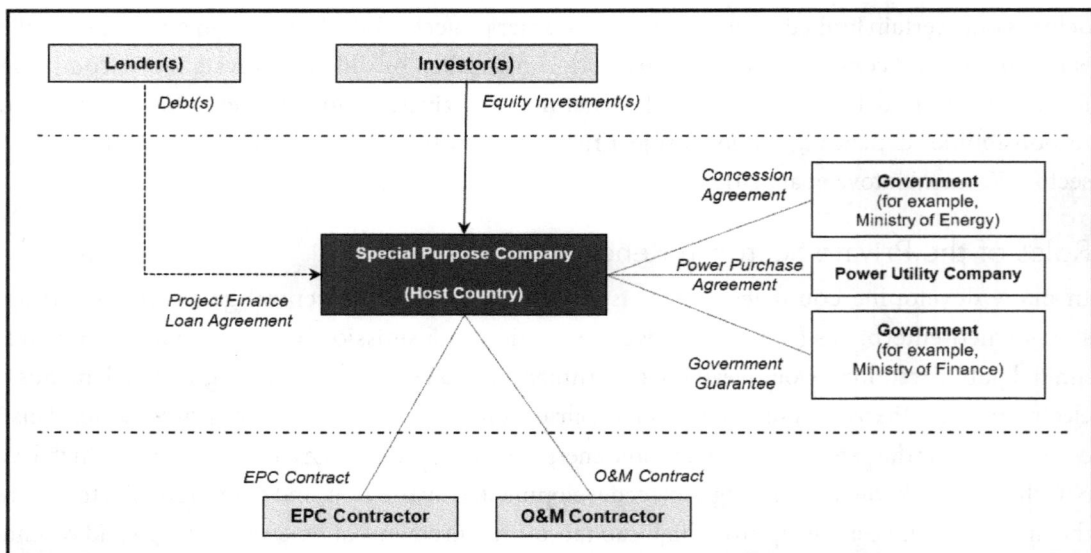

Figure 2.4.1 A Typical Renewable Energy Public-Private Partnership Project in a Developing Country

Inherent Risks in Renewable Energy Project

Unique risks that the government and investors usually face in the renewable energy PPP projects include, among others, resource quality, spatial planning, new technologies, administrative approval, local and environmental acceptance, and quality and price of maintenance services (Dinica, 2008). The distinctive characters of the renewable energy projects shape the unique features and risks of renewable energy PPP projects. Even though capital costs, that is, the up-front expense of building renewable energy projects during the construction stage, have dropped significantly since the early 2010s, renewable energy still poses a higher capital cost than other thermal energies. Despite low operation and maintenance costs, high capital costs usually make financial institutions perceive renewable energy projects as riskier, leading to a higher interest rate and requiring more stringent conditions to the investor. This fact is translated into the financial risks; therefore, the critical importance of robust financing mechanism needed in the renewable energy projects, where the financial gap should be innovatively bridged to foster the further development and wider diffusion of renewable energy technologies (Yue et al., 2001). High technical expertise and the management know-how required for project development and operations are associated with the construction and operation risks.

Additionally, the intermittent nature of renewable energy sources means that electricity is not continuously generated. It often cannot meet the peak energy demand if not combined with other energy storage solutions. Although the storing of energy may help ensure that power is available when needed most, it also entails higher capital costs, coupled with high operational and maintenance requirements, therefore posing additional financial risks and operational risks. Finally, economic incentives and policy support from the government, for example, agreed FITs, are often required to attract the renewable energy private sector developers by ensuring a stable revenue stream. They are stipulated in the PPA contract with a utility company, sometimes being guaranteed by the government. This leads to a risk of breach of contract by the government, a political risk in a project (Joskow & Tirole, 2005).

Private Sector Participants

Participation of a private sector company in a renewable energy project as an investor is usually driven by the investment return of a project on its equity capital measured by Return on Investment. The same private investor sometimes participates in the project as Engineering, Procurement, and Construction (EPC) contractor and/or Operation and Management (O&M) contractor to generate additional profits from EPC and O&M work. In addition to financial motivations, albeit low investment returns as an investor or low profitability in EPC and O&M, the private sector participates in renewable energy projects for other reasons such as market expansion, deployment and market test of new technology, and accumulation of track records for future projects (Bennett et al., 2000).

Specific knowledge-based skills and experiences are strongly required for the private sector participants in renewable PPP projects, and they are usually screened during the bid evaluation process. The project investors are expected to have strong technical and operational know-hows in the energy sector with financial strength and stability. EPC and O&M contractors are often required to have proven records and technical expertise in renewable energy projects. Given the complexity of financing arrangement, financial institutions should be able to make correct assessments of various risks involved in the projects, reflect them in the pricing, and analyze their impacts on the loan repayment.

The Public Sector

The public sector participants, either the government, SOE, or others, are also required to have specific skills and expertise, for example, a good understanding of the renewable energy technologies and trends, the ability to correctly assess and transfer certain risks to the private participants, evaluation of the price of the renewable energy project, negotiation with the power generators, effective and timely organization of community consultation, and monitoring capacity during the operation period. The lack of such required skills in the public sector, along with desperate needs for energy, may lead to poorly designed contracts that could, in turn, result in the unsuccessful implementation of renewable energy projects and/or ad-hoc unwanted renegotiations down the road (Martins et al., 2011). Furthermore, without knowledge of renewable energy technology and its market trends, the pricing may be determined at an arbitrary level, borne by the power users and eventually by taxpayers. Due to high visibility and its high level of environmental and social impacts on the community, the project may face difficulties in the future without proper consultation and community engagement processes.

The government's concerns are complex. It is required to be mindful of various dimensions of the project such as legislative and regulatory aspects, optimization of the proper energy mix, political opinions of the project, minimization of the retained risks and maximization of social benefits, and protection of taxpayers' interests. Nonetheless, the government has far more benefits to pursuing a PPP model in renewable energy generation in partnership with the private sector. The benefits include, among others, quick delivery of the project, learning up-to-date technological and managerial know-how, and harnessing financing capacities of the private sector (Jones, 1994). In return for these potential benefits, the public sector needs to show the private partners the stable revenue stream during the concession period to recoup their initial investments and make debt repayments.

Municipal and provincial governments are highly involved in most renewable energy projects, primarily due to environmental and social impacts on the local communities, such as noises and loss of visual landscape integrity. A local government's involvement includes identifying project sites, zoning plans, issuance of required licenses and permits, and local community consultation processes. Various forms of incentives, mostly tax incentives, are often provided from the central or upper-level to the local government to favor renewable energy development and accept a PPP renewable project in its province. Local governments generally support renewable energy projects because stable access to electricity is one of their key policy priorities to ensure residents' life quality and sustain a good business climate.

The following section introduces a PPP case for solar power generation in India and provides several lesson points for small-scale renewable PPP projects.

Case Study: India Gujarat Rooftop Solar Power Project in Gandhinagar

Case Overview

The State of Gujarat, located in western India with over 80 million residents, enjoys 300 sunny days per year. In recognizing the high solar energy potential, Gujarat province has embraced the idea of renewable energy to meet the state's growing energy demands. As Gujarat has been at the forefront of industrial development in India, the Gujarat government believed that renewable energy development would help sustain its significant national leadership in economic and social development while reducing the spread and depth of externalities and minimizing vulnerability to multiple spheres of economic growth. The push for renewable energy development was also motivated by the government's high recognition of the urgent need to diversify energy sources to preserve the state's endowed fossil fuels and to tackle climate change challenges simultaneously. The Gujarat government then adopted the Solar Power Policy of 2009 to install 500 MW solar power projects by 2014. The government also planned to make its capital, Gandhinagar, a pilot solar city, which can be instrumental in implementing the 2009 Solar Power Policy. To this end, the PPP-based rooftop solar photovoltaic (PV) technology was chosen as one of the strategic directions to achieve the policy goal, while effectively overcoming many technical, regulatory, and financial challenges regarding the solar rooftop projects (GEPD, 2009). With many other subsequent solar rooftop PPP projects, as of March 2020, the State of Gujarat was ranked the first in India for its domestic rooftop solar installation of 50,915 systems, which accounts for nearly 64% of the country's total of 79,950 systems. While this corresponds to about 177.67 MW power generation in the state each year, the installed rooftop solar systems' combined capacity is 322 MW, according to the Ministry of New and Renewable Energy.

Launched in 2010, the Gujarat rooftop solar power project in Gandhinagar is a landmark and pioneering small-scale project using solar PV technology under the 2009 Solar Power Policy framework. The project comprises the development, construction, and operation of two rooftop solar projects in the state capital city of Gandhinagar. The government sought private sector participation to finance and build two 2.5 MW pilot solar projects that could provide better access to power for an estimated 10,000 people (World Bank, 2012). For the project implementation, the Gujarat government granted private investors a concession to produce solar power of a total capacity of 5 MW. The generated power, in turn, has been purchased by Torrent Power, a private sector utility company under a 25-year PPA. The project went through a competitive bidding process based on a tariff that the bidders requested to quote. The world's 40 large energy technology firms had shown interest in solar projects. Four companies, Lanco Infratech, Azure Power, Sun Edison, and Mahindra Solar, were requested to submit the final bid. In 2011, the government selected Azure Power and Sun Edison, asking each company to install 2.5 MW solar generators. The project was commissioned in March 2012 (Rasika, 2012), and is now in full operation.

In this project, solar panels were installed on both public and private rooftops. The Gujarat government has provided the rooftops of about 25 state government buildings for around 4 MW power generation. The remaining 1 MW has been generated on 250 private properties, including apartment blocks, private houses, and commercial buildings. To ensure enough local communities' participation in the project, the government promised a "Green Incentive" of Rs. 3 (about $0.016 USD)/kWh to the participating city residents if they install solar panels on their rooftops. According to Rasika (2012), the monetary incentive was sufficient for inducing the residents' participation. It is projected that a rooftop owner would at best receive about Rs. 8,878 ($145) per month for a 1,000 sq. ft. of leased area, even with a very modest assumption of solar energy reception of about 200 KWh/sq. ft. per year and a conversion factor of 18%. This Green Incentive seemed to be innovative because the project, on the one hand, could be implemented without occupying and acquiring lands.

On the other hand, it also provides additional income sources to the participating residents for hosting the solar panels. The extra income was considered substantial, as the per capita income for 2009–2010 in Gujrat was Rs. 33,843 ($553) (GoG, 2013). Figure 2.4.2 shows the project structure of two Gujarat rooftop solar power projects of 2.5 MW capacity each.

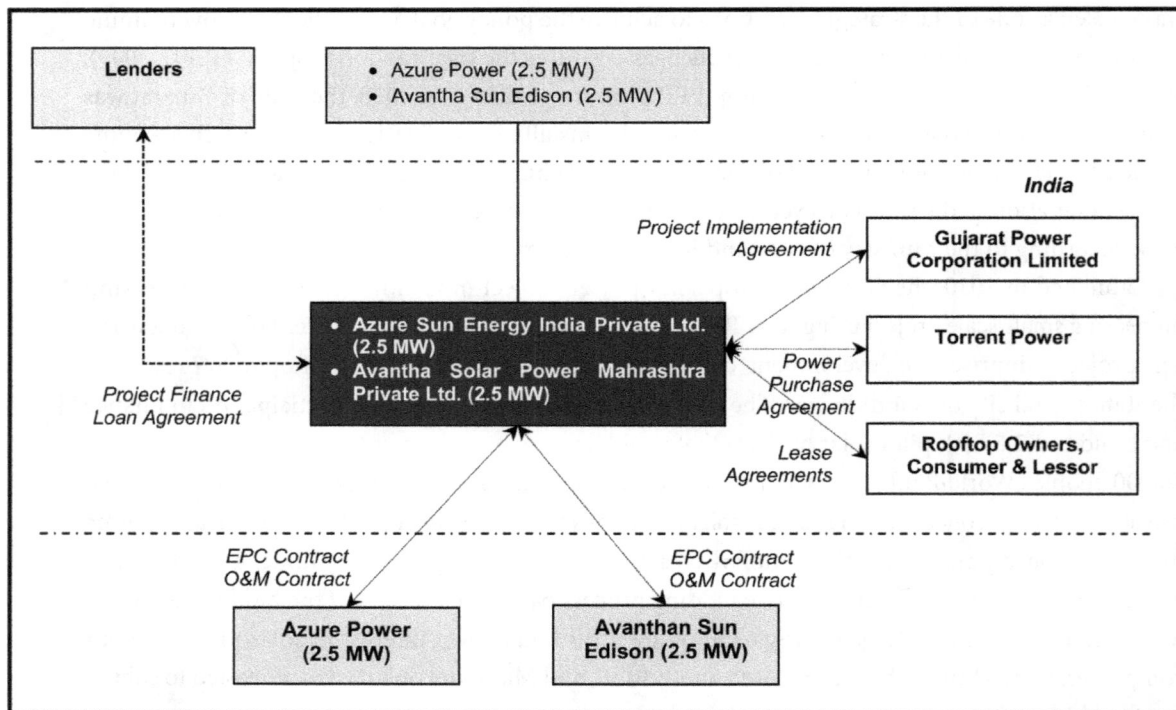

Figure 2.4.2 Project Structure of Gujarat Rooftop Solar Power.

The Gujarat government intended the project to prove the concept of small-scale grid-connected rooftop solar power development, which exists in developed countries (in the United States and Germany in particular). The project was structured to address the risks and challenges that constrain the rooftop solar market in India and thus provide path-finding solutions for market development (World Bank, 2015). For instance, securing an adequate number of suitable roofs (institutional, commercial, and residential) is a significant challenge for any distributed and grid-connected solar rooftop project. The government committed to leasing public building rooftops to mitigate the risk, while a green incentive mechanism was devised for residential buildings that can help monetize otherwise idle assets. The leasing of residential rooftops was also possible when targeted marketing was executed to ensure prospective customers who might have lacked the understanding of solar power generation have become well aware of the rights and obligations as lessors and the concept of rental payments. Off-taker was also identified and invited to discuss all technical, commercial, and legal aspects from the initial stage of the project development, which had resulted in the viability of the project. Other main challenges in the pilot project include lack of clear regulatory guidelines for interconnection of sub-1 MW generation facilities in the state electricity distribution code as well as lack of approval of the process and documents (such as PPA, project implementation agreement) by the regulators (World Bank, 2015). These were addressed during the early stage of the project before its bid process.

Overall, as the first MW-scale solar PPP project in India, the Gujarat project was considered very successful. The state government could enhance energy access at affordable prices with virtually no state subsidies, mobilize the private sector's investment and technical expertise, benefit the participating residents with additional rental incomes, and reduce carbon emissions (World Bank, 2012). The replicability of the same model in different cities adds another merit to the project. Relying on lessons learned from this pilot PPP model, the Gujarat government implemented subsequent 25 MW rooftop solar power projects. Gujarat Power Corporation Limited arranged the new projects in its five large cities, all in small scales, that is, Vadodara (5.0 MW), Rajkot (6.5 MW), Mehsana (5.0 MW), Bhavnagar (3.5 MW), and Surat (5 MW).[1]

Lessons Learned

There are important lessons drawn from the Gujarat Rooftop Solar Project for the renewable energy PPP projects in other developing countries. First, central and provincial governments' policy initiatives have been well coordinated to establish an enabling environment for renewable energy development, leading to the successful implementation of PPP projects in Gandhinagar. As part of India's National Action Plan for Climate Change, the Indian government launched Jawaharlal Nehru National Solar Mission (JNNSM) in 2010. The JNNSM has set its ambitious target of adding 1 GW of capacity between 2010 and 2013 and seeks to increase the combined solar power capacity from 9 MW in 2010 to 20 GW by 2022 (World Bank, 2010). One of the strategies to implement JNNSM was to bundle relatively expensive solar power with one from the unallocated quota of thermal power stations. The combined power generation approach was adopted to reduce the impact of a higher solar tariff on utilities. JNNSM also embraced a reverse bidding mechanism[2] that helped qualified bidders benefit from the declining global prices for solar components (World Bank, 2013). The Gujarat government's response to JNNSM was prompt and relied on the 2009 Solar Power Policy. The government implemented the state-specific regulations and guidelines for rooftop solar projects, adopting the "feed-in tariff" concept as a first-kind monetary incentive tool for India's renewable energy development (World Bank, 2015). It illustrates that PPP-based renewable projects would function well under the condition where a central government establishes a strategic direction for promoting renewable energy and implements its national strategy in harmony with state, provincial, or local governments' follow-up actions.

Second, the Gujarat rooftop pilot project indicates the significance of handling land constraints early in a project planning stage. Renewable energy projects usually face difficulties for site selection since it is mainly challenging to secure large parcels of lands where the right amount of natural resources, such as wind and sunlight, is continuously provided (Eliperin & Mufson, 2009). Besides, renewables have a low energy density compared to other conventional energy sources (Eliperin & Mufson, 2009). *Land acquisition* for solar PV or wind projects, regardless of its size, would be then translated into high up-front costs and greater environmental and social impacts about land conversion. In the Gujarat rooftop solar project case, private actors suggested the two rooftop-based solar power installments and the idea of utilizing existing local government buildings' rooftops. For the installment capacities fallen short of, residents were also invited for power generation. This way, being entirely accepted by the government led to the smooth implementation of the project while successfully lowering project costs and other resources that otherwise would be demanded land purchase negotiations.

Third, offering Green Incentive to project participants presents a win–win situation for both government and the local community. The Gujarat case is also called "Rent-a-Roof Project," as private companies lease rooftops from government and private residents who would receive Rs. 3 ($0.05) per unit produced. Solar panels are then installed and connected to the grid by project operators, and the operators, in turn, receive a FIT of Rs. 11.21 ($0.18) under a 25-year concession (World Bank, 2014). As such, this pilot project showcases a Green Incentive as a crucial design element for a compelling solar rooftop PPP business model. The incentive enables seamless collaborations among various stakeholders, including policymakers, regulators, individual rooftop owners (lease rentals), solar module suppliers, project developers, and utilities (meeting renewable purchase obligations by procuring solar power). Since the ability to pay for improved energy services is one of the major limitations in energy access problems in developing countries, Green Incentive, in sync with the preferred implementation model, would encourage clean energy usage after its installation. It would also prompt the scale-up of renewable-based electrification projects. Green Incentives can also enhance the local community's active involvement during a project operation stage that has been typically ignored in most renewable energy projects, despite its critical role in achieving overall project performance (Chaurey et al., 2012).

Following up on this pilot project's success, the Gujarat government has attempted to increase the rooftop solar model to other cities in the state. For example, the government sought a 5 MW rooftop solar project in Vadodara, and WAA Solar Private Limited financed around $8 million of the total project costs. The business model's slight modification was made to allow plants' ownership to be transferred to rooftop owners, which raised potential rooftop owners' interests in the project. With a 25-year PPA made, the project was successfully implemented and provided increased energy services to around 9,000 residents (World Bank, 2014). Likewise, solar rooftop PPPs with Green Incentives are replicable. As long as policy and regulatory frameworks are clear, stable enough to attract investors, and flexibility in PPP contracts are granted to project developers.

Conclusion

Energy security is one of the top policy priorities in a country, and climate change and its impacts have presented a twofold challenge of providing sustainable energy and divesting fossil fuel–oriented power generation. However, due to lack of financial means, limited technical expertise, and inadequate/insufficient regulatory mechanisms, developing countries would be more likely to supply sustainable energy with renewable projects. Those projects demand high up-front capital costs, establish regulatory guidelines and rules, and employ technical knowledge for project design and implementation.

PPP arrangements can be suggested as a procurement model to address such challenges. They can reduce risks from a capital-intensive cost structure, enable public actors to harness private sector's technical expertise, and encourage more effective project preparation (Martin et al., 2011). Key to the successful launch and undertaking of the project would be a careful consideration of design elements in PPP renewable projects' planning stage. The Gujarat Rooftop Solar Power project is a replicable PPP model for small-scale electrification projects in developing countries. The project's success would be attributed to multilevel governments' coordinated efforts for facilitating private investments by ways of incentive-based mechanisms such as lowered tariffs. Its uniqueness also includes avoiding unwanted negotiations over land acquisition for solar power generation, as the Gujarat project utilized the rooftops of public and private buildings. Green Incentive is a critical element that brings local residents' support for the project, inviting local residents to be part of power generators. Being replicated in other cities or states in India, the solar PPP model in Gandhinagar offers a promise for its replicability in other developing countries, insomuch as private companies make win–win operational arrangements with residents as well as a public sector create policy measures promoting renewable-based electricity generation.

While this chapter provides anecdotal evidence for crucial success factors in a solar PPP, a future scholarship can benefit from investigating PPP arrangements using different renewable sources such as wind. Also, it would be valuable to explore what inhibits or promotes the expansion of PPP-based renewable projects across other regions of the developing world, particularly in Africa suffering from the lowest electricity access rate in the world (IEA, 2016). In sum, this article suggests that well-designed renewable PPPs can be instrumental in improving energy access to meet rapidly growing domestic energy demands in developing countries. It would allow each country to participate in a global transition to a clean, low-carbon energy system to promote sustainable development.

Endnotes

1. Developing Asia is defined by the IEA to include Afghanistan, Bangladesh, Brunei, Cambodia, China, Chinese Taipei, DPR Korea, East Timor, India, Indonesia, Malaysia, Mongolia, Myanmar, Nepal, Pakistan, PDR Laos, Philippines, Singapore, Sri Lanka, Thailand, Vietnam, and Other Asia.

2. http://ppi-re.worldbank.org/~/media/GIAWB/RE/Documents/RE-data-all.xls

3. Biomass-based electricity generation projects were numbered 51 among 1,690 total financed projects from the World Bank PPRE Database.

4. http://www.gpclindia.com/showpage.aspx?contentid=110

5. The reverse auction as a competitive bidding mechanism allocates projects to a set of successful bidders with the lowest tariffs adding up to the quantum of capacity earmarked for each batch. JNNSM Phase I allowed benchmark tariff fixed for each financial year by the Central Electricity Regulatory Commission (CERC) (World Bank, 2013).

References

Abbott, K. (2012). Engaging the Public and the Private in Global Sustainability Governance. *International Affairs, 88*, 3.

Bennett, E., James, S., & Grohmann, P. (2000). Joint venture public partnerships for urban environmental services: report on UNDP/PPPUE's project development facility 1995–1999. United Nations Development Programme and Yale University.

Bridle R., & Kitson, L. (2014). The Impact of Fossil-fuel subsidies on renewable electricity generation. Global Subsidies Initiative Report. Geneva: International Institute for Sustainable Development. https://www.iisd.org/system/files/publications/impact-fossil-fuel-subsidies-renewable-electricity-generation.pdf

Chaurey, A., Krithika, P. R., Palit, D., Rakesh, S., & Sovacool, B. (2012). New partnerships and business models for facilitating energy access. *Energy Policy*, *47*, 48–55.

Dinica, V. (2008). Initiating a sustained diffusion of wind power: the role of public–private partnerships in Spain. *Energy Policy*, *36*, 3562–3571.

Economist. (2014, January 5). Why is renewable energy so expensive? http://www.economist.com/node/21592685

Eilperin, J., & Mufson, S. (2009, April 16). Renewable energy's environmental paradox. *The Washington Post.* http://www.washingtonpost.com/wpdyn/content/article/2009/04/15/AR2009041503622.html

Government of Gujarat (GoG). 2013. Socio-economic review. http://gujecostat.gujarat.gov.in/?page_id=12

Gujarat Energy and Petrochemicals Department (GEPD). (2009). Solar power policy. http://geda.gujarat.gov.in/policies_state.php

Gujarat Energy Development Agency (GEDA). (2015). Gujarat Electricity Regulatory Commission (GERC) order determination of tariff for solar energy projects. http://geda.gujarat.gov.in/policies_state.php

Grégor Q., Jarrousse, A., & Mouen, S. (2013). Developing renewable energies in Africa: A public-private partnership. *Private Sector and Development*, 25–27. http://www.proparco.fr/webdav/site/proparco/shared/PORTAILS/Secteur_prive_developpement/PDF/SPD18/SPD18_Gregor_Quiniou_Astrid_Jarrousse_Stephanie_Mouen_UK.pdf

Indian Ministry of New and Renewable Energy (IMNRE). (2016). Solar power capacity milestone of 5,000 MW in India. http://pib.nic.in/newsite/printrelease.aspx?relid=134497

IEA Renewable Energy Working Party. (2002). Renewable energy into the mainstream, 9. http://anetce.com/2002_iea_renewables54.pdf

IFC. (2014). India: Gujarat Solar. Public private partnership in series. http://www.ifc.org/wps/wcm/connect/d0a75c804b077348b4acfe888d4159f8/PPPStories_India_GujaratSolar.pdf?MOD=AJPERES

IFC. (2014). Harnessing energy from the sun: Empowering rooftop owners. https://www.ifc.org/wps/wcm/connect/topics_ext_content/ifc_external_corporate_site/sustainability-at-ifc/publications/p_report_harnessing_energy_from_the_sun

International Energy Agency (IEA). (2016). World energy outlook. http://www.worldenergyoutlook.org/resources/energydevelopment/energyaccessdatabase/

IPCC. (2014). Climate Change 2014: Synthesis report. https://www.ipcc.ch/report/ar5/https://www.ipcc.ch/report/ar5/

Jacob, A. (2005). Wind power market shows no signs of slowing. *Reinforced Plastics*, *49*, 24–29.

Jones, B. D. (1994). *Reconceiving decision-making in democratic politics: Attention, choice, and public policy.* The University of Chicago Press.

Joskow, P. L., & Tirole, J. (2005). Merchant transmission investment. *Journal of Industrial Economics*, *53*(2), 233–264.

Koch, C., & Buser, M. (2006). Emerging meta-governance as an institutional framework for public private partnerships networks in Denmark. *International Journal of Project Management*, *24*, 548–556.

Komendantova, N., Patt, A., Barras, L., & Battaglini, A. (2012). Perception of risks in renewable energy projects: The case of concentrated solar power in North Africa. *Energy Policy*, 40,103–109.

Komendantova, N., Patt, Anthony, & Williges, Keith. (2011). Solar power investment in North Africa: Reducing perceived risks. *Renewable and Sustainable Energy Reviews*, *15*(9), 4829–4835.

Komor, P. (2004). *Renewable energy policy.* iUniverse Inc.

Kruckenberg, L. J. (2015). Renewable energy partnerships in development cooperation: Towards a relational understanding of technical assistance. *Energy Policy*, *77*, 11–20.

Martins, A. C., Marques, R. C., Cruz, C. O. (2011). Public-private partnerships for wind power: The Portuguese case. *Energy Policy*, *39*, 94–104.

Mittai, S. (2014). 750 MW solar power plant in India, likely to be largest solar power plant in world, gets World Bank financing commitment. *Clean Technica*. Retrieved May 1, 2017, from http://cleantechnica .com/2014/12/20/largest-solar-power-plant-world-750-mw-solar-power-plant-india-gets-world-bank -financing-commitment/

Pessoa, A. (2010). Reviewing public-private partnership performance in developing economies. In G. A. Hodge, G. Carsten, & A. E. Boardman (Eds.), *International handbook on public-private partnerships* (pp. 568–593). Edward Elgar Publishers.

Pattberg, P. H. (2010). Public-Private partnerships in global climate governance. *Climate Change*, *1*(2), 279–287.

Pattberg, P. H., Biermann, F., Chan S., & Mert, A. (2012). Public-private partnerships for sustainable development: Emergence, influence, and legitimacy needs? *Review of African Political Economy*, *40*(137), 485–495.

Rasika, G. A. (2012). India's PPP model for rooftop solar programme. www.energetica-india.net

Renewable Energy Policy Network for the 21st Century (REN 21). (2005). Renewables 2005: Global status report. http://www.ren21.net/wp-content/uploads/2016/06/GSR_2016_Full_Report.pdf

REN 21. (2016). Renewables 2016: Global status report. http://www.ren21.net/Resources/Publications/ REN21Publications.aspx

Sovacool, B. K. (2011). *Developing public-private renewable energy partnerships to expand energy access.* Report for the United Nations Economic and Social Commission for the Asia Pacific, Bangkok, Thailand. Institute for Energy and the Environment, Vermont Law School.

Sovacool, B. K. (2013). Expanding renewable energy access with pro-poor public private partnerships in the developing world. *Energy Strategy Reviews*, *1*, 181–192.

UNDP. (2011). Public private partnerships for service delivery (PPPSD), UNDP capacity development group, Johannesburg, South Africa.

UNEP. (2017). Global trends in renewable energy investment. http://fs-unep-centre.org/sites/default/files/ publications/globaltrendsinrenewableenergyinvestment2017.pdf

UNFCCC. (2016). Paris Agreement: Report of the conference of the parties on its twenty-first session. http:// unfccc.int/meetings/paris_nov_2015/items/9445.php

World Bank. (2010). Unleashing the potential of renewable energy in India. http://documents.worldbank.org/ curated/en/504181468260121463/Unleashing-the-potential-of-renewable-energy-in-India

World Bank. (2012). *India: Gujarat solar. Public–private partnerships brief.* World Bank Group. http://documents
.worldbank.org/curated/en/2015/06/24581474/india-gujarat-solar

World Bank. (2013). Paving the way for a transformational future: Lessons from Jawaharlal Nehru National
Solar Mission Phase I. Energy Sector Management Assistance Program. https://openknowledge.worldbank
.org/handle/10986/17480

World Bank. (2014). Replicating success in Vadodara: Rooftop solar PPPs in India. http://www.worldbank.org/
en/news/feature/2014/09/25/replicating-success-in-vadodara-rooftop-solar-ppps-in-india

World Bank. (2015). Rooftop solar public-private partnerships: Lessons from Gujarat solar. https://library
.pppknowledgelab.org/documents/2408

World Economic Forum. (2013). Scaling up energy access through cross-sector partnerships. http://www3
.weforum.org/docs/IP/2014/EN/Communications/WEF_EN_Scaling_Up_Energy_Access.pdf

WHO. (2009). The energy access situation in developing countries. http:// www.who.int/indoorair/publications/
energyaccesssituation/en/

Yue, C., Liu, C., & Liou, E. (2001). A transition toward a sustainable energy future: Feasibility assessment and
development strategies of wind power in Taiwan. *Energy Policy, 29,* 951–963.

DISCUSSION QUESTIONS

1. What are the pros and cons of renewable energy PPPs from the government's, private investor's, and financier's perspectives?

2. What are the prominent uniqueness, benefits, and challenges of the rooftop solar PPP projects, compared to other solar PPP projects, as evidenced by the Gujarat case study?

3. What are the other noticeable trends of renewable energy projects in developing countries that you see from the database in https://ourworldindata.org/? Suppose penetration of a certain renewable technology in the electricity mix visibly increases in Sub-Saharan African countries during the last 5–7 years. What is that technology, and what do you think are the reasons?

Carbon Markets

Principles and Current Practices

By Younsung Kim

Introduction

Acid rain was known to change surface water chemistry, lead to declines in fish populations, reduce forest growth, increase plant disease, and accelerate materials damage. Scientific research and case studies confirmed that acid rain was attributed to sulfur dioxide (SO_2) and nitrogen oxide (NO_x)—emissions from anthropogenic sources, primarily industrial processes and utilities (Cowling, 1982). In response, the United States government developed the first national cap-and-trade program by establishing the SO_2 allowance trading program. (Layzer & Rinfret, 2020). Allowance trading programs or cap-and-trade systems are approaches that harness market forces to reduce emissions cost-effectively. In an allowance trading program, a government sets an emissions cap and issues a quantity of emission allowances consistent with that cap. Companies that can reduce their emissions at a lower cost may sell any excess allowances to companies that would not meet the emissions cap cost-effectively (Grubb, 2014).

In 1995 the SO_2 allowance trading program was established under the Title IV Acid Deposition Control of the Clean Air Act Amendments of 1990, requiring major emission reductions of SO_2 and NO_x from the power sector (Layzer & Rinfret, 2020; United States Environmental Protection Agency [EPA], 2022). It proved to be highly effective in achieving emission reductions, reducing emissions quickly with an overall lower cost than the mandatory regulatory approach (Stavins, 2007). After a twenty-five-year operation, the program ended with the final 2010 SO_2 cap set at 8.95 million tons, a level of about one-half of the emissions from the power sector in 1980 (EPA, 2022). The success of the SO_2 allowance trading program was modeled and led to the adoption of emissions trading programs for greenhouse gases (GHGs) to mitigate climate change.

This article reviews how carbon markets have evolved in the context of international climate governance. The European Union Emissions Trading System (EU ETS), the first CO_2 allowance trading program, is used to showcase the development of a carbon market in which each installment meets its carbon target flexibly (see Figure 2.5.1). It also documents the status and trends of carbon

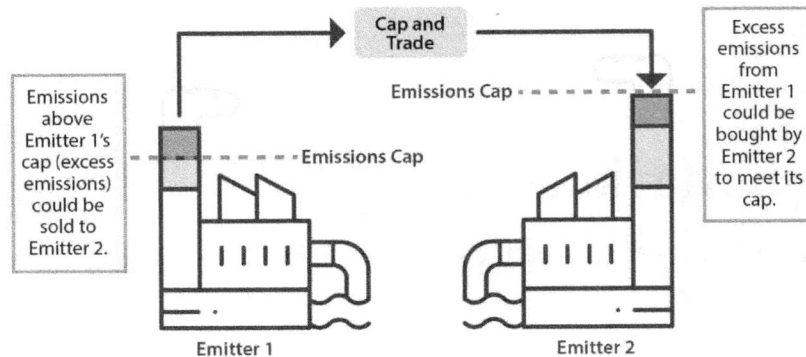

Figure 2.5.1 Basic function of an emissions trading program.

markets, drawing attention to the continuing efforts to expand global carbon markets and create carbon pricing mechanisms that could profoundly affect individuals' production and consumption choices and contribute to a carbon-neutral world by 2050.

Cap-and-Trade Approach

According to neoclassical economists, markets can distribute limited resources efficiently. However, a market can fail for several reasons ranging from monopoly, information asymmetries, and externalities to public goods. A market failure is a situation where there is an inefficient allocation of goods and services in a free market. When it comes to the environment, market failure occurs for two main reasons. First, environmental goods are public goods, and markets are less likely to provide environmental quality or greater environmental performance without government incentives. Second, a market system fails due to negative externalities. An externality refers to a cost or benefit resulting from a transaction that affects a third party. It can be negative or positive. Industrial pollution is a classic example of a negative externality that creates uncompensated costs to third parties, while a positive externality is considered unpaid benefits to third parties. Vaccination and public education are examples of positive externalities. Suppose there is a community adjacent to a fossil fuel-based power plant. If residents in the community experience serious health issues caused by air pollutants from the power plant but their health-related costs are not fully compensated, a negative externality arises.

An allowance trading program can help address negative externalities. It forces polluters to internalize their production or consumption costs external to third parties. Within a cap-and-trade program, regulators distribute permits to companies and allow them to trade emissions reduced beyond their emissions targets. Companies that cannot meet their targets have choices to either reduce emissions internally or buy permits from the emissions trading market. Through this approach, overall emissions are reduced, but the costs for emission reductions are substantially lower than the costs under mandatory regulations. In mandatory regulations, all firms are required to reduce emissions up to their targets regardless of their pollution abatement capacities (Olmstead, 2022). Therefore, reducing emissions via a market-based system allows regulated sources to select the cost-effective option for regulatory compliance. The cap-and-trade program also provides a compelling reason for investing in innovating technologies and practices for further emissions reduction. If companies successfully reduce carbon emissions and have additional carbon permits that can be sold in a carbon market, they can generate revenues from the additional carbon emissions reduced. In Figure 2.5.2, Company A needs to reduce 50 tCO_{2e} from 200 tCO_{2e} to 150 tCO_{2e}. In order to meet this goal, it can buy 50 tCO_{2e} from Company B, which has already met its cap of 100 tCO_{2e}. Company B can further reduce its emissions and sell 50 tCO_{2e} to Company A as long as the additional carbon reduction is profitable. Thus, both companies are able to achieve their carbon targets.

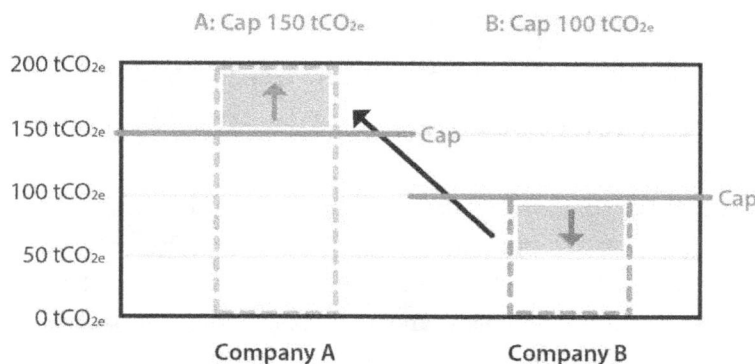

Figure 2.5.2 Carbon permit trade between two companies in a hypothetical cap-and-trade program.

The Kyoto Protocol and Carbon Markets

Carbon markets have arisen from the trade of carbon offset credits, which are market-based instruments that assign a monetary price to carbon. They are driven by international climate agreements designed to address climate change and are integral to global climate governance. With mounting evidence on climate science, the United Nations Framework Convention on Climate Change (UNFCCC) was established in 1992 and agreed to stabilize atmospheric greenhouse gas concentrations. With this aim, countries decided to meet at the Conference of Parties (COP) to discuss their carbon control progress. In 1997, parties at the third COP meeting adopted the Kyoto Protocol, the first legally binding international climate treaty. The Kyoto Protocol aimed to reduce 5.2 percent of greenhouse gases from the 1990 baseline over a period between 2008 to 2012. It is important to note that the emissions reduction targets and obligations were only applied to thirty-eight developed nations (described as "Annex I"), exempting developing nations (Selin & VanDeveer, 2022). In doing so, the protocol followed the principle of common but differentiated responsibilities (CBDR), under which all states have a shared obligation to address environmental destruction, but a country's commitment target was set up differently depending on its historical contributions to climate change.

Countries were allowed to meet the Kyoto targets by relying on domestic carbon reduction policies such as energy efficiency increases in their economies, carbon sinks, and Kyoto Mechanisms. The Kyoto Mechanisms are considered flexible and are divided into project-based and allowance-based emissions saving or emissions reduction. Clean Development Mechanism (CDM) and Joint Implementation (JI) are project-based, lowering GHG emissions by projects in non-Annex I countries and in the party nations, respectively. The projects that count as emissions reduction range from building renewable-based utility farms, foresting lands, and capturing methane from industrial livestock facilities. The European Union used allowance-based carbon reductions, creating the EU ETS to meet the EU's Kyoto emission target.

EU Emissions Trading System (ETS) and US Regional Greenhouse Gas Initiative (RGGI)

The EU ETS, the largest and oldest carbon market, demonstrates that international emissions trading for GHGs is feasible. It enabled companies to reduce emissions wherever they would be cheapest, not limited to national borders. Under the EU ETS, more than 40 percent of the EU's emissions were capped. The EU ETS was designed initially among the fifteen West European Member States and automatically extended to the twelve countries of Eastern Europe that joined the EU between 2005 to 2007. Three neighboring countries, Norway, Liechtenstein, and Iceland, also subsequently joined, so that in 2022 it spanned thirty-one countries and included more than 114,000 plants. Table 2.5.1 shows national or regional ETSs that have been implemented since 2005.

The first phase of EU ETS from 2005 to 2007 was called the learning by trial period, and the following period, from 2007 to 2012, was in alignment with the First Commitment Period of the Kyoto Protocol. In this setting, the EU ETS was legally coupled with the Kyoto Protocol, and exchanges of EU allowances across borders were matched by transfers of national Assigned Amount Units. This linkage was extended further by coupling the EU ETS to other emissions offset mechanisms, mostly the Clean Development Mechanism. Linking emissions trading schemes to the Kyoto Protocol's flexible mechanisms could lead to a decrease in carbon prices and a reduction of the overall compliance costs (Nazifi, 2010). The ETS Linking Directive allowed credits from such other mechanisms to be purchased by EU installations to offset their emission caps (Grubb, 2014).

There were three key features of phase 1 (2005–2007). First, it covered only CO_2 emissions from power generators and energy-intensive industries. Second, almost all allowances were given to businesses for free. Third, the penalty for noncompliance was €40 per tonne. While a carbon market was developed, there were noticeable fallouts during the first phase. Most member states had no system

Table 2.5.1 National or regional ETS implemented.

Name of the Initiative	Type of Jurisdiction Covered	Jurisdiction Covered	Year of Implementation
Canada Federal OBPS	National	Canada	2019
China national ETS	National	China	2021
EU ETS	National	EU, Norway, Iceland, Liechtenstein	2005
Germany ETS	National	Germany	2021
Kazakhstan ETS	National	Kazakhstan	2013
Korea ETS	National	Korea, Republic of	2015
Mexican Pilot ETS	National	Mexico	2020
New Zealand ETS	National	New Zealand	2008
Switzerland ETS	National	Switzerland	2008
UK ETS	National	United Kingdom	2021

Data source: Carbon pricing dashboard (n.d.).

in place to measure GHG emissions from industry, which were left to make their own emissions projections, leading to an overestimate of emissions by some sectors, notably German utilities. This resulted in the price crashing in 2007 at the end of Phase 1 (see Figure 2.5.3).

In response, the cap in the second phase was set 6.5 percent lower compared to 2005, which increased the EU allowance price to around €30 in 2008. Nitrous oxide emissions were also included by a number of countries, and free allocation was reduced to around 90 percent. The reform of the ETS framework for phase 3 (2013–2020) changed the system considerably from phases 1 and 2. The main changes included a single, EU-wide cap on emissions, auctioning for allowance allocations, and the inclusion of additional sectors and gases, such as aviation and nitrous oxide emissions from the production of nitric acid included by a number of countries (European Commission, n.d.).

As explained in the price collapse, the EU ETS was not proven effective in phase 1. The EU ETS did not cause reductions (Petrick & Wagner, 2014), and the CO_2 price showed no impact on emissions

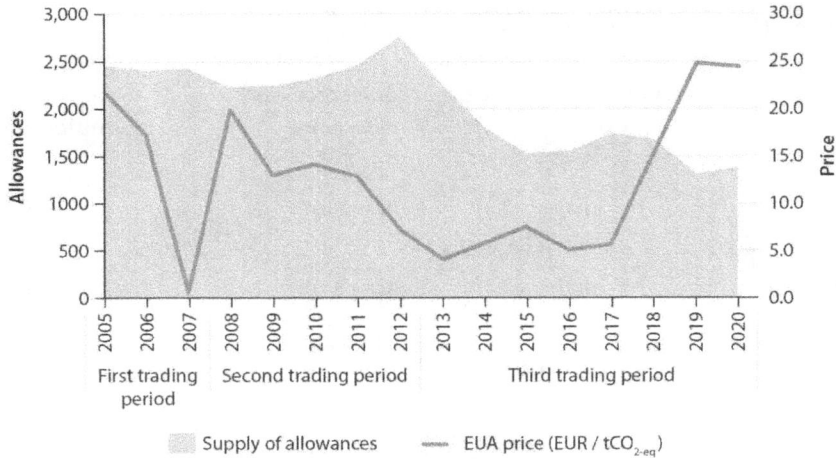

Figure 2.5.3 Price versus supply of EU allowances (EUA), 2005–2020.
Source: Adapted from European Environment Agency (2022).

(Gloaguen & Alberola, 2013). However, phase 2 showed substantive carbon reductions, drastically increasing trading volumes compared to phase 1 (see Figure 2.5.4). German manufacturing firms reduced their emissions between 25 percent and 28 percent from 2008 to 2010, as compared to unregulated firms. French manufacturing firms reduced emissions between 13.5 percent and 19.8 percent during the same time period, largely due to fuel switching (Wagner et al., 2014).

While many lessons were learned from the first two phases, the second phase's longevity developed confidence in a successful system, marking the EU ETS as a primary policy instrument for reducing

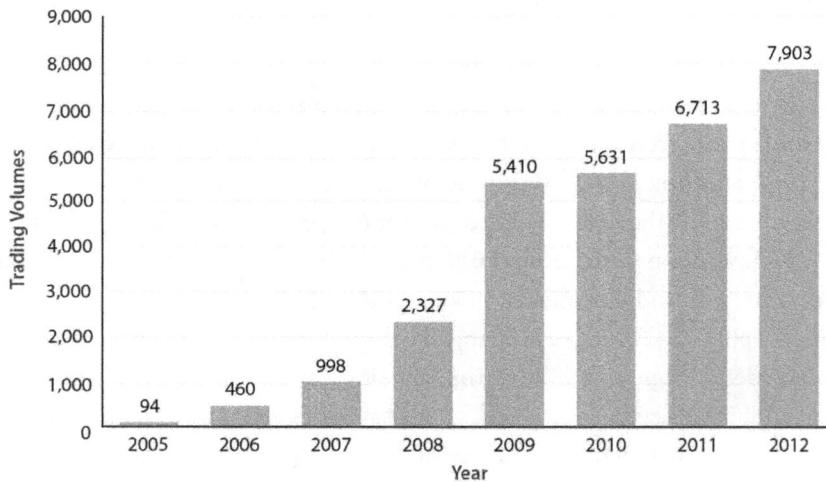

Figure 2.5.4 Trading volumes in EU emission allowances from 2005 to 2012 (in billions).
Source: Adapted from European Commission (n.d.).

greenhouse gases in the EU. The EU ETS is now in its fourth phase (2021–2030), and the EU has raised carbon reduction goals to 40 percent for GHG reduction below 1990 levels, and 27 percent for both renewable energy generation and improved energy savings by 2030 (Selin & VanDeveer, 2022).

The US Regional Greenhouse Gas Initiative (RGGI) can be pointed out as another carbon market example, established similarly to EU ETS. It started as a regional cap-and-trade system among electricity generators in nine US northeastern states, covering about 18 percent of regional emissions in 2020. The RGGI was a positive force to fill a void of the US federal climate inaction when the adoption of a national economy-wide carbon emission trading failed under the Obama administration (Kim, 2022). RGGI has been regarded as successful toward reducing CO_2 emissions within participating states, with estimates of 9 million tons of reductions per year. However, it has been counterbalanced by emission increases in surrounding states, leading to reductions of about one-half of estimates (Fell & Maniloff, 2018).

Efforts to Sharpen and Broaden the Carbon Market at the COP 26 Meeting

Recently, more governments are adopting net-zero targets as well as ambitious carbon pricing instruments via either emissions trading systems or carbon taxes (see Table 2.5.2). In the EU, carbon allowance prices have hit all-time highs, representing continually growing demands for carbon credits in the carbon market. The EU carbon permit price in 2022 was €78 (Trading Economics, 2022), while a carbon permit in the NZ emissions trading scheme was being traded at NZ $76 (Carbon News, 2022). As of 2021, twenty-eight subnational, regional, and national emissions trading systems had been implemented, covering around 16 percent of annual greenhouse gas emissions (World Bank, 2022).

Increasing carbon price programs and expanding carbon markets after the Paris climate deal is a cheerful victory, but several concerns about the market-based carbon reductions have remained. In general, the economic incentive-based emissions trading was criticized by some environmentalists as giving leeway to polluters who may avoid emissions reductions and instead simply buy out carbon permits to meet their compliance targets. Also, the actual effects of the emissions reduction policy are known to be limited, as carbon prices are not high enough to generate substantial emissions reductions (Green, 2021). Double counting was a particular concern related to the technical aspect of emissions reduction accounting. Double counting occurs when two countries claim the same carbon removal or emission reduction. For instance, if a developing nation reduced its carbon emissions by one metric ton through a solar-power program, it might be tempted to sell a reduction credit to a developed country while also counting the reduction in its own target (Kizzier et al., 2019). This situation also leads to a relocation of emissions rather than a net reduction, creating leakage. When leakage occurs, emissions reductions may be overestimated (Green, 2021).

Table 2.5.2 National or regional carbon tax implemented.

Name of the Initiative	Type of Jurisdiction Covered	Jurisdiction Covered	Year of Implementation
Argentina carbon tax	National	Argentina	2018
Canada federal fuel charge	National	Canada	2019
Chile carbon tax	National	Chile	2017
Colombia carbon tax	National	Colombia	2017
Denmark carbon tax	National	Denmark	1992
Estonia carbon tax	National	Estonia	2000
Finland carbon tax	National	Finland	1990
France carbon tax	National	France	2014
Iceland carbon tax	National	Iceland	2010
Ireland carbon tax	National	Ireland	2010
Japan carbon tax	National	Japan	2012
Latvia carbon tax	National	Latvia	2004
Liechtenstein carbon tax	National	Liechtenstein	2008
Luxembourg carbon tax	National	Luxembourg	2021
Mexico carbon tax	National	Mexico	2014
Netherlands carbon tax	National	Netherlands	2021
Norway carbon tax	National	Norway	1991
Poland carbon tax	National	Poland	1990
Portugal carbon tax	National	Portugal	2015
Singapore carbon tax	National	Singapore	2019
Slovenia carbon tax	National	Slovenia	1996
South Africa carbon tax	National	South Africa	2019
Spain carbon tax	National	Spain	2014
Sweden carbon tax	National	Sweden	1991
Switzerland carbon tax	National	Switzerland	2008
UK Carbon Price Support	National	United Kingdom	2013
Ukraine carbon tax	National	Ukraine	2011
Uruguay CO_2 tax	National	Uruguay	2022

Data source: Carbon pricing dashboard (n.d.).

This concern has been one of the main agendas at the COP 26 meeting in Glasgow in 2022. As an achievement of the meeting, Article 6 of the Paris Agreement related to carbon market rules was revised to provide clear accounting guidelines for emissions credits in carbon markets. According to the revision, the country that generates a credit will decide whether to authorize it for sale to other nations or to count it toward their own climate targets. If authorized and sold, the seller country will add an emission unit to its national tally, and the buyer country will deduct one, to ensure the emissions cut is counted only once between countries (Brooks & Adler, 2021).

Figure 2.5.4 COP 26 climate summit in Glasgow, Scotland, November 2021. The COP meetings bring together world leaders and organizers to discuss and negotiate climate initiatives, some regarding carbon markets.
Photo courtesy Ian Forsyth, Getty Images.

The COP 26 decisions also led to high expectations for the accelerated expansion of voluntary carbon markets that are formed by increased investments in carbon credits and demand for such credits. The carbon demand growth will then boost the value of carbon credits, which will help to get more ambitious emission reduction projects started. The potential market size of international carbon credits is assumed to be between $100–400 billion per year by 2030, enabling governments and businesses pledging voluntary carbon targets to reach their targets (Brooks & Adler, 2021). COP 26 rightly called on countries to reduce global carbon emissions by 45 percent by 2030 to achieve net-zero carbon dioxide emissions by 2050 (Kizzier, 2021).

As such, carbon markets will play a critical role in meeting country pledges toward a 1.5°C temperature reduction goal under the Paris Agreement. Carbon markets can facilitate the phasedown of greenhouse gas emissions at an accelerating speed and the needed energy transition to renewables. Carbon markets will also contribute to the scaled-up realization of sustainable development. Ultimately, individuals under countries and companies striving to achieve carbon neutrality by mid-century are poised to bear the costs for their carbon emissions to address climate change.

Bibliography

Bodle, R., Donat, L., & Duwe, M. (2016). The Paris Agreement analysis, assessment and outlook. *Carbon Climate Law Review, 10*, 1. https://www.ecologic.eu/sites/default/files/event/2016/ecologic_institu...

Brooks, C., & Adler, K. (2021, November 8). COP26: Article 6 rulebook updated but remains work in progress. *S&P Global.* https://www.spglobal.com/commodityinsights/en/market-insights/latest-ne...

Carbon News. (2022). *Carbon prices.* https://www.carbonnews.co.nz/tag.asp?tag=Carbon+prices

Cowling, E. B. (1982). Acid precipitation in historical perspective. *Environmental Science & Technology, 16*, 110A–123A. https://doi.org/10.1021/es00096a725

European Commission. (n.d.). *Development of EU ETS (2005–2020).* https://ec.europa.eu/clima/eu-action/eu-emissions-trading-system-eu-ets...

European Commission. (2013). *The EU Emissions trading system (EU ETS).* https://climate.ec.europa.eu/eu-action/eu-emissions-trading-system-eu-e...

European Environment Agency. (2022, January 5). *Emissions, allowances, surplus and prices in the EU ETS, 2005–2020.* https://www.eea.europa.eu/data-andmaps/figures/emissions-allowances-su...

Fell, H., & Maniloff, P. (2018). Leakage in regional environmental policy: The case of the regional greenhouse gas initiative. *Journal of Environmental Economics and Management, 87*, 1–23. https://doi.org/10.1016/j.jeem.2017.10.007

Gloaguen, O., & Alberola, E. (2013). *Assessing the factors behind CO2 emissions changes over the phases 1 and 2 of the EU ETS: An econometric analysis* (Working Paper). CDC Climate Research. https://inis.iaea.org/search/search.aspx?orig_q=RN:45063760

Green, J. F. (2021). Does carbon pricing reduce emissions? A review of ex-post analyses. *Environmental Research Letters, 16*(4), 43004. https://doi.org/10.1088/1748-9326/abdae9

Grubb, M. (2014). *Energy, climate change and the three domains of sustainable development.* Routledge.

Kim, Y. (2022). Integrated market and nonmarket strategies: Empirical evidence from the S&P 500 firms' climate strategies. *Business and Politics, 24*, 1–22. https://doi.org/10.1017/bap.2021.18

Kizzier, K. (2021, November 13). COP26 ends with a strong result on carbon markets and an international call to action for the most urgent climate priorities. *Environmental Defense Fund.* https://www.edf.org/media/cop26-ends-strong-result-carbon-markets-and-i....

Kizzier, K., Levin, K., & Rambharos, M. (2019, December 2). What you need to know about Article 6 of the Paris Agreement. *World Resources Institute*. https://www.wri.org/insights/what-you-need-know-about-article-6-paris-a....

Lake, K. (2013, May 1). Learning from Europe's carbon price crash: We need a carbon bank. *The Conversation*. https://theconversation.com/learning-from-europes-carbon-price-crash-we...

Layzer, J. A., & Rinfret, S. R. (2020). *The environmental case: Translating values into policy* (5th edition). CQ Press.

Nazifi, F. (2010). The price impacts of linking the European Union emissions trading scheme to the clean development mechanism. *Environmental Economics and Policy Studies, 12*, 164–186. https://doi.org/10.1007/s10018-010-0168-3

Olmstead, S. (2022). Applying market principles to environmental policy, In N. J. Vig, M. E. Kraft, & B. G. Rabe (Eds.), *Environmental policy: New directions for the twenty-first century*. CQ Press.

Petrick, S., & Wagner, U. J. (2014). *The impact of carbon trading on industry: Evidence from German manufacturing firms* (Kiel Working Paper No. 1912). Kiel Institute for the World Economy. https://doi.org/10.2139/ssrn.2389800

Selin, H., & VanDeveer, S. D. (2022). Global climate change governance: Can the promise of Paris be realized? In N. J. Vig, M. E. Kraft, & B. G. Rabe (Eds.), *Environmental policy: New directions for the twenty-first century*. CQ Press.

Stavins, R. (2007). Market-based environmental policies: What can we learn from U.S. experience (and related research)? In J. Freeman & C. D. Kolstad (Eds.), *Moving to markets in environmental regulation* (pp. 19–47). Oxford University Press.

Trading Economics. (2022). *EU carbon permits*. https://tradingeconomics.com/commodity/carbon

United States Environmental Protection Agency. (2022). *Acid rain program*. https://www.epa.gov/acidrain/acid-rain-program

Wagner, U. J., Muuls, M., Martin, R., & Comer, J. (2014, June 28–July 2). *The causal effects of the European Union Emissions Trading Scheme: Evidence from French Manufacturing Plants* [Conference presentation]. Fifth World Congress of Environmental and Resource Economics, Istanbul, Turkey. http://conference.iza.org/conference_files/EnvEmpl2014/martin_r7617.pdf

World Bank. (2021, May 25). State and trends of carbon pricing 2021. *The World Bank Group eLibrary*. https://doi.org/10.1596/978-1-4648-1728-1

World Bank. (2022). *Carbon pricing dashboard*. https://carbonpricingdashboard.worldbank.org/map_data

Wråke, M., Burtraw, D., Löfgren, A., & Zetterberg, L. (2012). What have we learnt from the European Union's Emissions Trading System? *Ambio, 41*(Suppl 1), 12–22. https://doi.org/10.1007/s13280-011-0237-2

DISCUSSION QUESTIONS

1. Carbon markets are based on the principle of the cap-and-trade program when reducing greenhouse gases. As compared to carbon taxes or other prescriptive climate policy measures, what are the benefits? And what are the limitations of cap-and-trade programs?

2. In creating the EU Emissions Trading System (ETS), the EU designed the first phase from 2005 to 2007 as a learning process to understand whether the market system works. Explain the characteristics of the first phase of the EU ETS and discuss how they guided the development of other emissions trading systems in different institutional contexts.

3. Voluntary carbon markets have emerged as a policy tool to offset hard-to-decarbonize emissions for corporate climate and net-zero strategies. What are the challenges and potential of voluntary carbon markets? What urgent tasks could international policymakers and governments do to leverage voluntary carbon markets for countries' carbon neutrality policy goals?

www.ingramcontent.com/pod-product-compliance
Lightning Source LLC
Chambersburg PA
CBHW081538220326
41598CB00036B/6482